Alexej Schelle

**Formation of a dilute Bose-Einstein condensate**

AF092545

Alexej Schelle

# Formation of a dilute Bose-Einstein condensate

## Number-conserving master equation theory of Bose-Einstein condensation

Südwestdeutscher Verlag für Hochschulschriften

**Impressum/Imprint (nur für Deutschland/ only for Germany)**
Bibliografische Information der Deutschen Nationalbibliothek: Die Deutsche Nationalbibliothek verzeichnet diese Publikation in der Deutschen Nationalbibliografie; detaillierte bibliografische Daten sind im Internet über http://dnb.d-nb.de abrufbar.

Alle in diesem Buch genannten Marken und Produktnamen unterliegen warenzeichen-, markenoder patentrechtlichem Schutz bzw. sind Warenzeichen oder eingetragene Warenzeichen der jeweiligen Inhaber. Die Wiedergabe von Marken, Produktnamen, Gebrauchsnamen, Handelsnamen, Warenbezeichnungen u.s.w. in diesem Werk berechtigt auch ohne besondere Kennzeichnung nicht zu der Annahme, dass solche Namen im Sinne der Warenzeichen- und Markenschutzgesetzgebung als frei zu betrachten wären und daher von jedermann benutzt werden dürften.

Verlag: Südwestdeutscher Verlag für Hochschulschriften Aktiengesellschaft & Co. KG
Dudweiler Landstr. 99, 66123 Saarbrücken, Deutschland
Telefon +49 681 37 20 271-1, Telefax +49 681 37 20 271-0
Email: info@svh-verlag.de
Zugl.: Univ. Pierre et Marie Curie, Albert-Ludwigs Univ. Freiburg, binat. Diss., 2009

Herstellung in Deutschland:
Schaltungsdienst Lange o.H.G., Berlin
Books on Demand GmbH, Norderstedt
Reha GmbH, Saarbrücken
Amazon Distribution GmbH, Leipzig
**ISBN: 978-3-8381-1289-3**


**Imprint (only for USA, GB)**
Bibliographic information published by the Deutsche Nationalbibliothek: The Deutsche Nationalbibliothek lists this publication in the Deutsche Nationalbibliografie; detailed bibliographic data are available in the Internet at http://dnb.d-nb.de.

Any brand names and product names mentioned in this book are subject to trademark, brand or patent protection and are trademarks or registered trademarks of their respective holders. The use of brand names, product names, common names, trade names, product descriptions etc. even without a particular marking in this works is in no way to be construed to mean that such names may be regarded as unrestricted in respect of trademark and brand protection legislation and could thus be used by anyone.

Publisher: Südwestdeutscher Verlag für Hochschulschriften Aktiengesellschaft & Co. KG
Dudweiler Landstr. 99, 66123 Saarbrücken, Germany
Phone +49 681 37 20 271-1, Fax +49 681 37 20 271-0
Email: info@svh-verlag.de

Printed in the U.S.A.
Printed in the U.K. by (see last page)
**ISBN: 978-3-8381-1289-3**


Copyright © 2010 by the author and Südwestdeutscher Verlag für Hochschulschriften Aktiengesellschaft & Co. KG and licensors
All rights reserved. Saarbrücken 2010

*Gewidmet meinem Vater, sowie unseren Großvätern Paul Proksch und Josef Schelle.*

# Formation of a dilute Bose-Einstein condensate: Number-conserving master equation theory of condensate formation

$$\varrho_\perp \lambda^3 \geq \zeta(3/2)$$

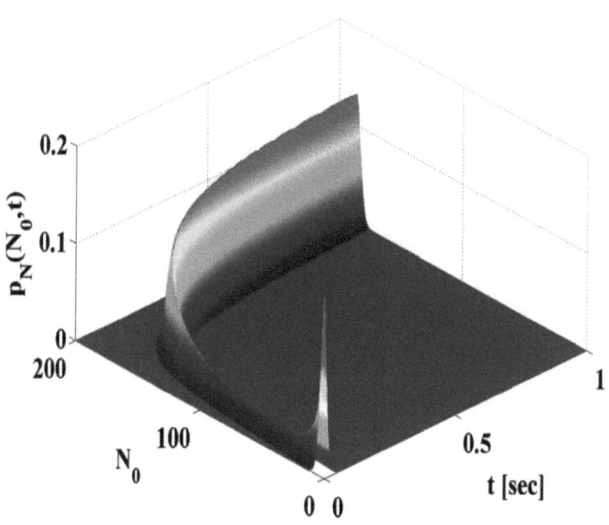

The figure displays the time evolution of the condensate number distribution during Bose-Einstein condensation after a quench of the Bose gas' temperature below the critical temperature expected for condensate formation.

*Ob eine Stadt zivilisiert ist, hängt nicht von der Zahl ihrer Schnellstrassen ab, sondern davon, ob ein Kind auf dem Dreirad unbeschwert überall hin kommt.*

Enrique Penalosa, August, 2008

6

# Contents

**Introduction to the thesis**    1
    Motivation of this thesis . . . . . . . . . . . . . . . . . . . . . . . . 1
    How to model Bose-Einstein condensation microscopically? . . . . . . . . . 3
    Outline of the thesis . . . . . . . . . . . . . . . . . . . . . . . . . . 4

**I  CONCEPTS OF ULTRACOLD MATTER THEORY**    9

**1  Bose-Einstein condensation in ideal Bose gases**    19
    1.1  What is a Bose-Einstein condensate? . . . . . . . . . . . . . . . . 19
    1.2  What is quantum ergodicity? . . . . . . . . . . . . . . . . . . . 20
    1.3  Original prediction of Bose-Einstein condensation . . . . . . . . . . 21
    1.4  Experimental state-of-the-art . . . . . . . . . . . . . . . . . . 26
    1.5  Bose-Einstein condensation in harmonic traps . . . . . . . . . . . . 28
        1.5.1  Grand canonical ensemble . . . . . . . . . . . . . . . . 29
        1.5.2  The canonical ensemble . . . . . . . . . . . . . . . . . 36
    1.6  Bose-Einstein condensation in position space . . . . . . . . . . . . 38

**2  Interacting Bose-Einstein condensates**    41
    2.1  S-wave scattering approximation . . . . . . . . . . . . . . . . . 41
    2.2  Hamiltonian for two body interactions . . . . . . . . . . . . . . 44
    2.3  Gross-Pitaevskii equation from the Hartree ansatz . . . . . . . . . 45

| | | | |
|---|---|---|---|
| 2.4 | Theories of condensate growth | | 48 |
| | 2.4.1 | Condensate growth from quantum Boltzmann equation | 48 |
| | 2.4.2 | Pioneering works of Levich and Yakhot | 51 |
| | 2.4.3 | Predictions of Kagan, Svistunov and Shlyapnikov | 51 |
| | 2.4.4 | Kinetic evolution obtained from Holland, Williams and Cooper | 53 |
| | 2.4.5 | Stoof's contribution | 54 |
| | 2.4.6 | Quantum kinetic theory | 54 |
| Survey: Which current aspects can we adopt to monitor the many body dynamics during Bose-Einstein condensation? | | | 60 |

## II  QUANTUM MASTER EQUATION OF BOSE-EINSTEIN CONDENSATION — 63

### 3  Concepts, basic assumptions and validity range — 67

| | | | |
|---|---|---|---|
| 3.1 | Motivation for master equation: Separation of time scales | | 68 |
| 3.2 | Modeling of many particle dynamics | | 69 |
| | 3.2.1 | Two body interactions in dilute gases | 70 |
| | 3.2.2 | Condensate and non-condensate subsystems | 70 |
| | 3.2.3 | Thermalization in the non-condensate | 71 |
| 3.3 | $N$-body Born-Markov ansatz | | 73 |
| | 3.3.1 | General Born-Markov ansatz | 73 |
| | 3.3.2 | Born ansatz for gases of fixed particle number | 74 |
| | 3.3.3 | Markov approximation for a Bose-Einstein condensate | 75 |
| 3.4 | Limiting cases and validity range | | 76 |
| | 3.4.1 | Dilute gas condition | 76 |
| | 3.4.2 | Perturbative limit | 77 |
| | 3.4.3 | Thermodynamic limit | 77 |
| | 3.4.4 | Semiclassical limit | 78 |
| | 3.4.5 | Physical realization of limiting cases | 78 |

| | | | |
|---|---|---|---|
| **4** | **Quantized fields, two body interactions and Hilbert space** | | **81** |
| | 4.1 | Definition of the condensate | 82 |
| | 4.2 | Interactions between condensate and non-condensate | 83 |
| | | 4.2.1 Separation of the second quantized field | 83 |
| | | 4.2.2 Decomposition of the Hamiltonian | 84 |
| | | 4.2.3 Two body interaction processes | 85 |
| | 4.3 | Hamiltonian of the non-condensate background gas | 89 |
| | | 4.3.1 Diagonalization of the non-condensate Hamiltonian | 90 |
| | | 4.3.2 Perturbative spectrum of non-condensate particles | 93 |
| | 4.4 | Hilbert spaces | 95 |
| | | 4.4.1 Single particle Hilbert space | 95 |
| | | 4.4.2 Fock-Hilbert space | 95 |
| | | 4.4.3 Fock-Hilbert space of states with fixed particle number | 96 |
| **5** | **Lindblad master equation for a Bose-Einstein condensate** | | **99** |
| | 5.1 | Evolution equation of the total density matrix | 99 |
| | 5.2 | Time evolution of the reduced condensate density matrix | 102 |
| | | 5.2.1 $N$-body Born ansatz | 102 |
| | | 5.2.2 Evolution equation for the condensate | 104 |
| | 5.3 | Contribution of first order interaction terms | 105 |
| | | 5.3.1 General operator averages in the Bose state | 105 |
| | | 5.3.2 Vanishing of linear interaction terms | 106 |
| | 5.4 | Dynamical separation of two body interaction terms | 108 |
| | 5.5 | Lindblad operators and transition rates | 109 |
| | | 5.5.1 Lindblad evolution term for single particle processes ($\leadsto$) | 112 |
| | | 5.5.2 Lindblad evolution term for pair processes ($\leftrightsquigarrow$) | 116 |
| | | 5.5.3 Evolution term for scattering processes ($\circlearrowright$) | 120 |
| | 5.6 | Quantum master equation of Lindblad type | 121 |

## III  Environment-induced dynamics in Bose-Einstein condensates  125

### 6  Monitoring the Bose-Einstein phase transition  129
  6.1  Dynamical equations for Bose-Einstein condensation .......... 130
    6.1.1  Master equation of Bose-Einstein condensation ......... 131
    6.1.2  Growth equations for average condensate occupation ..... 134
    6.1.3  Condensate particle number fluctuations ............. 136
  6.2  Bose-Einstein condensation in harmonic traps ................ 137
    6.2.1  Monitoring of the condensate number distribution ....... 138
    6.2.2  Dynamics of the condensate number variance .......... 141
    6.2.3  Average condensate growth from the thermal cloud ...... 143
  6.3  Comparison of formation times to state-of-the-art ............ 146

### 7  Transiton rates for Bose-Einstein condensation  149
  7.1  Single particle (⤳), pair (⬌) and scattering (⟲) rates .......... 149
    7.1.1  Single particle feeding and loss rate ................ 149
    7.1.2  Pair feeding and loss rates ....................... 153
    7.1.3  Two body scattering rates ....................... 155
  7.2  Depletion of the non-condensate ......................... 156
  7.3  Detailed particle balance conditions ....................... 157
  7.4  Single particle, pair and scattering energy shifts ............. 158
  7.5  Transition rates and energy shifts in the perturbative limit ...... 162
    7.5.1  Leading order of transition rates ................... 163
    7.5.2  Leading order energy shifts ...................... 168
  7.6  Generalized Einstein de Broglie condition ................... 169

### 8  Equilibrium properties of a dilute Bose-Einstein condensate  173
  8.1  Equilibrium steady state after Bose-Einstein condensation ..... 173
  8.2  On the quantum ergodicity conjecture .................... 175
  8.3  Exact condensate statistics versus semiclassical limit .......... 179
    8.3.1  Condensate particle number distribution ............. 179

|   |   | 8.3.2 | Average condensate occupation and number variance . . . . . | 183 |
|---|---|---|---|---|

       8.3.2   Average condensate occupation and number variance . . . . . 183
       8.3.3   Shift of the critical temperature . . . . . . . . . . . . . . . . . . . . . . 184
  8.4   Analytical scaling behaviors in the semiclassical limit . . . . . . . . . . 185
       8.4.1   Condensate and non-condensate particle number distribution 185
       8.4.2   Average condensate occupation and number variance . . . . . 188
       8.4.3   Higher order moments of the steady state distribution . . . . 189

**9  Final conclusions    193**
  9.1   Master equation of Bose-Einstein condensation . . . . . . . . . . . . . . 193
  9.2   What is Bose-Einstein condensation? . . . . . . . . . . . . . . . . . . . . 194
  9.3   Outlook . . . . . . . . . . . . . . . . . . . . . . . . . . . . . . . . . . . . . . 195

**Appendix    199**

**A  Important proofs and calculations    199**
  A.1   Correlation functions of the non-condensate field . . . . . . . . . . . . 199
  A.2   Detailed balance conditions . . . . . . . . . . . . . . . . . . . . . . . . . 203
  A.3   Occupation numbers of the non-condensate . . . . . . . . . . . . . . . . 204
  A.4   Proof of uniqueness of the Bose gas' steady state . . . . . . . . . . . . 207
  A.5   Non-condensate thermalization . . . . . . . . . . . . . . . . . . . . . . . 209

**Bibliography    211**

# Introduction

**Motivation of this thesis**

Bose-Einstein condensates open the path for the in situ investigation of several interesting many particle effects in atomic gases such as superfluidity [1, 2], or quantized vortices [3, 4, 5]. Due to the coherent wave nature of ultracold quantum matter, Bose-Einstein condensates are in particular perfectly suited to study a vast range of quantum phenomena based on quantum coherence – like Anderson localization [6, 7, 8], or Josephson oscillations [9] – on the micrometer scale. Latter scenarios are usually known from other fields of physics, such as the theory of quantum optics, or the realm of solid state theory.

Besides these wonderful examples how to manipulate and employ Bose-Einstein condensates with high precision in order to access these different physical branches in present days' experiments, there remains a fundamental theoretical question concerning the condensate formation process: How can we describe the quantum dynamics of the Bose-Einstein phase transition beyond the evolution of the average macroscopic ground state occupation, connecting the experimental observations of average macroscopic occupation with a dynamical, microscopic many particle picture?

Another motivation of the present work arises from the applicational point of view. The parameter regime of typical state of the art experiments does in principle not match the validity range in which fundamental thermodynamical postulates [10], leading to the thermal state ansatz for the equilibrium state of the quantum gas, can be taken for granted. Since the theory of thermodynamics is supposed to be valid only for total particle numbers of the order of Avogadro's number, $N \sim 10^{23}$, fundamental assumptions [11] like equipartition of energy (ergodicity) and the existence of a unique and stable equilibrium state are strictly justified only in the thermodynamic limit, under the neglect of number and energy uncertainties. These assumptions may not be realistic for Bose-Einstein condensates in the quantum degenerate limit, because they consist of a few thousands of atoms [12] – thus being far from the thermodynamic limit.

Moreover, the atoms in a Bose-Einstein condensate interact via (species) specific collision processes, the occupation numbers of the different energy modes fluctuate, and the particles exhibit strong phase coherences due to their indisputably quantum mechanical nature at low temperatures.[1] How is it possible, as conjectured by the thermodynamics of ideal gases, that a quantum gas will always follow the statistics of a Boltzmann thermal state of non-interacting particles in the limit of weak (but non-zero) interactions, independently of their type, i.e. lacking any hysteresis on the condensate formation process?

These reflections demonstrate that the so called "Boltzmann ergodicity conjecture" [13, 14], originating from classical, statistical mechanics, is nontrivial, especially for weakly interacting quantum systems of finite size. Under which conditions does a Bose-Einstein condensate exhibit a unique and stable equilibrium steady state – and, how can we characterize such state in analytical terms? Is the statistics and the dynamics of a Bose-Einstein condensate well described by an ideal gas, if the atomic sample is sufficiently dilute? If yes, to which extend does the specific type of the atomic interactions play a role for the microscopic many particle dynamics of Bose-Einstein condensation? And how does the finite particle number of a Bose-Einstein condensate influence the condensate dynamics and the Boltzmann equilibrium statistics of the gas?

In summary, we hence address two essential issues in the present thesis:

⋄ Our dynamical, microscopic understanding of the Bose-Einstein phase tran-

[1] where the wave length of the particles is of the order of their average distance

sition, in particular concerning the interplay of particle number fluctuations below the critical temperature for Bose-Einstein condensation, $T_c$, and the creation of this new state of matter – the Bose-Einstein condensate – is so far incomplete. How can we link the microscopic many particle dynamics during the Bose-Einstein phase transition to the buildup of a macroscopically occupied ground state mode? Which role plays the particle-wave duality, what is the impact of interactions and finally which role plays spatial quantum coherence of the bosonic atoms for the process of Bose-Einstein condensation?

⋄ The answer to the question [11] whether a dilute, weakly interacting Bose-Einstein condensate exhibits a unique and stable equilibrium steady state. How close and under which assumptions does a Bose-Einstein condensate, consisting of a finite number of weakly interacting atoms – as given in realistic state-of-the-art experiments [12] – follow the Gibbs-Boltzmann statistics of an ideal gas? To which extent are finite size effects, quantum fluctuations and particle-particle interactions essential for the condensate equilibrium statistics?

## How to model Bose-Einstein condensation microscopically?

Answers to these questions require a direct way of modeling the quantum many particle dynamics of the Bose gas, i.e., a theory beyond the mean field ansatz mostly studied in the literature [15, 16].

We generally consider the derivation of a master equation [17, 18, 19, 20, 21] as one of the most efficient and powerful tools to study Bose-Einstein condensation. To this end, we use the separation of time scales between the rapid non-condensate thermalization dynamics from the comparably slow condensate formation time, considering the condensate as a system part which evolves in time under the dynamically depleted thermal non-condensate environment. Deriving the master equation, we hence (i) account for all two body particle-particle interactions, (ii) circumvent a factorization of the $N$-body state of the gas into a condensate and non-condensate density matrix, (iii) assume particle number conservation, and (iv) take into account the depletion of the non-condensate thermal component during condensate formation.

Employing these experimental conditions for a quantum gas in our master equation formalism leads to a fundamentally new master equation ansatz which provides in particular experimentally desired condensate formation rates, through the first dynamical monitoring of the condensate and non-condensate particle number distributions during condensate formation. Arising condensate number fluctuations garnish the onset of the condensate formation process below $T_c$, until they reduce after the approach towards a steady state.

The master equation's stationary solution defines this equilibrium steady state for the $N$-body state of the gas under the inclusion of the wave nature of the quantum particles below $T_c$, number fluctuations and weak two body interactions. This enables the comparison of a microscopically derived equilibrium steady state of a dilute, weakly interacting Bose-Einstein condensate with a Gibbs-Boltzmann thermal state of exactly $N$ non-interacting, indistinguishable particles.

The physical bottom line of our theory is the first direct monitoring of condensate *and* non-condensate particle number distributions during condensate formation. This is based upon the connection of two fundamental properties, *particle number conservation* and *rapid non-condensate thermalization*, to extent the conventional Born-Markov ansatz to the $N$-body state of the gas of fixed particle number. This $N$-body Born-Markov ansatz together with the capitalization of the dilute gas condition $a\varrho^{1/3} \ll 1$ reduce the complex dynamics of the Bose-Einstein phase transition to a numerically accessible quantum master equation.

## Outline of the thesis

In *Part I* of the thesis, the most important state-of-the-art concepts for treating Bose-Einstein condensates are summarized.

Starting from Einstein's original prediction of Bose-Einstein condensation for non-interacting, uniform gases in **Chapter I**, theoretical extensions to the case of external confinements are discussed. We explain how Bose-Einstein condensates are currently created in state-of-the-art experiments, and deduce a perturbative parameter for our theory, characterizing the dilute gas regime. Care is taken to point out discrepancies between the grand canonical and canonical ensemble for condensate statistics of indistinguishable particles below the critical temperature, which persist even in the thermodynamic limit.

In **Chapter II**, the s-wave scattering approximation relating to two body interactions in dilute atomic gases is explained, and the concept of second quantized bosonic fields is introduced. We sketch the derivation of the Gross-Pitaevskii equation for the condensate wave function in terms of the Hartree ansatz, and summarize the existing theories for the study of average condensate growth.

*Part II* is dedicated to the development of a Lindblad quantum master equation theory of Bose-Einstein condensation.

The conceptual ingredients of the quantum master equation theory are summarized in **Chapter III**, focussing on the Markovian dynamics assumption, on two body interactions, on the constraint of particle number conservation and on the description of the non-condensate depletion during condensate formation, required for the derivation of the master equation in Chapter IV. We explain the validity range of the quantum master equation theory which applies to dilute atomic gases.

In **Chapter IV**, we start with the microscopic description for the Bose gas in second quantization, through the definition of the condensate and the non-condensate. This naturally provides a decomposition of the many particle Hamiltonian for dilute atomic gases, which allows us to derive a Lindblad quantum master equation for the condensate degree of freedom under the Markov dynamics assumption, with the nontrivial part of the dynamics induced by two body interaction processes. To do so, there is particular need to analyze the underlying single particle Hilbert space of wave functions, and the many particle Fock-Hilbert space structure.

In **Chapter V**, the Lindblad master equation for the time evolution of the reduced condensate density matrix is derived, describing the time evolution of the entire state of the Bose gas. The Lindblad master equation yields formal expressions for all transition rates and energy shifts associated with two body collision processes between condensate and non-condensate atoms.

In *Part III*, we employ the Lindblad quantum master equation to understand the quantum mechanical characteristics of the Bose-Einstein phase transition numerically and analytically. Evolution equations describing Bose-Einstein condensation are extracted, yielding in particular time scales for condensate formation. The equilibrium steady state of the gas of $N$ bosonic particles is harvested from the Lindblad quantum master equation.

We hence first extract the master equation for the diagonal elements of the condensate density matrix from the Lindblad master equation for practical purposes in **Chapter VII**. This allows us to numerically study the time evolution of condensate and non-condensate occupation numbers during condensate formation, and to extract the dynamical behavior of quantum matter fluctuations during Bose-Einstein condensation. We compare condensate formation times to previous theoretical predictions and to experimental observations.

In **Chapter VI**, we show how the formally defined transition rates and associated energy shifts are evaluated within a perturbative approach for the condensate wave function, valid for dilute and weakly interacting gases. Explicit analytical expressions for transition rates and energy shifts in a three-dimensional harmonic trap are obtained. We derive balance conditions for the transition rates, and deduce a generalized Einstein de Broglie condition for Bose-Einstein condensation.

Finally in **Chapter VIII**, it is proven that the steady state solution of the master equation defines a unique and stable equilibrium steady state of the Bose gas. We proof analytically and verify numerically that this steady state is a Gibbs-Boltzmann thermal state of an ideal gas within the Markovian dynamics assumption and in the limit of weak interactions. We oppose the steady state to predictions in the semiclassical limit, and deduce the shift of the critical temperature. Explicit analytical expressions for all moments of the condensate particle number distribution valid in the limit of large atomic gases complete the analysis of the present thesis.

**Chapter IX** concludes the conceptual and physical results of the present work and formulates some open questions and perspectives.

# Part I

# CONCEPTS OF ULTRACOLD MATTER THEORY

*"Not everything that counts can be counted, and not everything that can be counted counts."*

Albert Einstein

12

CODATA 2006 [22]

| Physical constant | Symbol | Numerical value | Unit |
|---|---|---|---|
| Speed of light | $c$ | $2.99792458 \times 10^8$ | s$^{-1}$ |
| | | $2.99792458 \times 10^{10}$ | cm s$^{-1}$ |
| Planck constant | $h$ | $6.62606896(33) \times 10^{-34}$ | J s |
| | | $6.62606896(33) \times 10^{-27}$ | erg s |
| | $hc$ | $1.239841875(31) \times 10^{-6}$ eV m | |
| Planck constant/$2\pi$ | $\hbar$ | $1.054571628(53) \times 10^{-34}$ | J s |
| | | $1.054571628(53) \times 10^{-27}$ | erg s |
| Elementary charge | $e$ | $1.602176487(40) \times 10^{-19}$ | C |
| Electron mass | $m_e$ | $9.10938215(45) \times 10^{-31}$ | kg |
| | | $9.10938215(45) \times 10^{-28}$ | kg |
| | $m_e c^2$ | $0.510998910(13)$ | MeV |
| Proton mass | $m_p$ | $1.672621637(83) \times 10^{-27}$ | kg |
| | $m_p c^2$ | $938.272013(23)$ | MeV |
| Atomic mass unit | $m(C^{12})/12$ | $1.660538782(83) \times 10^{-27}$ | kg |
| | $m_u c^2$ | $31.494028(23)$ | MeV |
| Boltzmann constant | $k_B$ | $1.3806504(24) \times 10^{-23}$ | J K$^{-1}$ |
| | | $1.3806504(24) \times 10^{-16}$ | erg K$^{-1}$ |
| | | $8.617343(15) \times 10^{-5}$ | eV K$^{-1}$ |
| | $k_B/h$ | $2.0836644(36) \times 10^{10}$ | Hz K$^{-1}$ |
| | | $20.836644(36)$ | Hz nK$^{-1}$ |
| Fine structure constant | $\alpha_f^{-1}$ | $137.035999679(94)$ | |
| Bohr radius | $a_B$ | $5.2917720859(36) \times 10^{-11}$ | m |
| Classical electron radius | $\frac{e^2}{4\pi\epsilon_0 m_e c^2}$ | $2.8179402894(58) \times 10^{-15}$ | m |
| Atomic unit of energy | $\frac{e^2}{4\pi\epsilon_0 a_0}$ | $27.21138386(68)$ | eV |

## NOTATION GUIDE

### Latin Letters

| | | |
|---|---|---|
| $\hat{a}_k, \hat{a}_k^\dagger$ | particle annihilation and creation operators | Eq. (4.5) |
| $a$ | s-wave scattering length | Eq. (2.6) |
| $\mathscr{F}$ | Fock–Hilbert space | Eq. (4.37) |
| $\mathscr{F}_0$ | condensate Fock–Hilbert space | Eq. (4.37) |
| $\mathscr{F}_\perp$ | non-condensate Fock–Hilbert space | Eq. (4.37) |
| $\mathscr{F}(N)$ | Fock–Hilbert space of $N$ particles | Eq. (4.40) |
| $\mathscr{F}(N-N_0)$ | non-condensate Fock–Hilbert space of $(N-N_0)$ non-condensate particles | Eq. (4.39) |
| $\mathscr{G}_\rightsquigarrow^{(\pm)}(\vec{r},\vec{r}',N-N_0,T,\tau)$ | normal (+) and anti-normal (−) correlation function for single particle processes $\rightsquigarrow$ | Eqs. (5.32, 5.33) |
| $\mathscr{G}_\leftrightsquigarrow^{(\pm)}(\vec{r},\vec{r}',N-N_0,T,\tau)$ | normal (+) and anti-normal (−) correlation function for pair processes ($\leftrightsquigarrow$) | Eqs. (5.48, 5.49) |
| $\mathscr{G}_\circlearrowleft(\vec{r},\vec{r}',N-N_0,T,\tau)$ | correlation function for scattering processes ($\circlearrowleft$) | Eq. (5.59) |
| $h_1$ | first quantized single particle Hamiltonian | Eq. (1.13) |
| $\hat{\mathcal{H}}$ | second quantized Hamiltonian of the gas | Eq. (4.7) |
| $\hat{\mathcal{H}}_0$ | second quantized condensate Hamiltonian | Eq. (4.8) |
| $\hat{\mathcal{H}}_\perp$ | second quantized non-condensate Hamiltonian | Eq. (4.9) |
| $H_{l_x}, H_{l_y}, H_{l_z}$ | hermite polynomials | Eq. (7.31) |
| $\mathscr{H}$ | single particle Hilbert space | Eq. (4.35) |
| $\mathscr{H}_0$ | condensate single particle Hilbert space | Eq. (4.35) |
| $\mathscr{H}_\perp$ | non-condensate single particle Hilbert space | Eq. (4.35) |
| $g = 4\pi a\hbar^2/m$ | two body interaction strength | Eq. (2.6) |
| $\mathscr{L}_\rightsquigarrow$ | Lindblad superoperator for single particle events ($\rightsquigarrow$) | Eq. (5.42) |
| $\mathscr{L}_\leftrightsquigarrow$ | Lindblad superoperator for pair events ($\leftrightsquigarrow$) | Eq. (5.53) |
| $L_x, L_y, L_z$ | harmonic oscillator lengths in $x, y$ and $z$ direction | Eq. (1.14) |
| $m_\alpha[z]$ | Bose function | Eq. (7.48) |
| $f_k(N-N_0,T)$ | average single particle occupation numbers of the non-condensate with $(N-N_0)$ particles | Eq. (7.7) |
| $\hat{\mathscr{P}}^\pm(N_0)$ | pair quantum jump operators | Eq. (5.44) |
| $p_N(N_0,T)$ | condensate particle number distribution of master equation | Eq. (5.11) |
| $p_N(N-N_0,T)$ | non-condensate particle number distribution of master equation | Eq. (5.11) |
| $\hat{\mathscr{P}}^\pm(N_0)$ | single particle quantum jump operators | Eq. (5.43) |

## NOTATION GUIDE

### Latin Letters

| | | |
|---|---|---|
| $T_c$ | critical temperature of the Bose gas | Eq. (1.5) |
| $\hat{U}(t)$ | time evolution operator with respect to $\hat{\mathcal{H}}$ | Eq. (5.6) |
| $\hat{U}_0(t)$ | time evolution operator with respect to $\hat{\mathcal{H}}_0$ | Eq. (5.6) |
| $\hat{U}_\perp(t)$ | time evolution operator with respect to $\hat{\mathcal{H}}_\perp$ | Eq. (5.6) |
| $\hat{V}_{0\perp}$ | condensate and non-condensate interactions | Eq. (4.10) |
| $\hat{V}_{\rightsquigarrow}$ | single particle interactions ($\rightsquigarrow$) | Eq. (4.16) |
| $\hat{V}_{\leftrightsquigarrow}$ | pair interactions ($\leftrightsquigarrow$) | Eq. (4.14) |
| $\hat{V}_{\circlearrowleft}$ | scattering interactions ($\circlearrowleft$) | Eq. (4.15) |
| $z$ | fugacity | Eq. (1.18) |
| $\mathscr{Z}_{GC}(\mu, T)$ | grand canonical partition sum | Eq. (1.17) |
| $\mathscr{Z}_C(N, T)$ | canonical partition sum | Eq. (1.26) |

### Greek Letters, Labels

| | | |
|---|---|---|
| $\rightsquigarrow$ | single particle events $\Delta N_0 = -\Delta N_\perp = \pm 1$ | Eq. (4.10) |
| $\leftrightsquigarrow$ | pair events $\Delta N_0 = -\Delta N_\perp = \pm 2$ | Eq. (4.10) |
| $\circlearrowleft$ | scattering events $\Delta N_0 = \Delta N_\perp = 0$ | Eq. (4.10) |
| $\hat{\gamma}_k, \hat{\gamma}_k^\dagger$ | particle operators associated to the modes $|\Theta_k\rangle$ | Eq. (4.20) |
| $\Delta_{\rightsquigarrow}^{(\pm)}(N-N_0, T)$ | energy shift for single particle events ($\rightsquigarrow$) | Eq. (5.41) |
| $\Delta_{\leftrightsquigarrow}^{(\pm)}(N-N_0, T)$ | energy shift for pair events ($\leftrightsquigarrow$) | Eq. (5.52) |
| $\Delta_{\circlearrowleft}^{(\pm)}(N-N_0, T)$ | energy shift for scattering events ($\rightsquigarrow$) | Eq. (5.62) |
| $\epsilon_k$ | eigenenergies of non-condensate single particle states $|\Psi_k\rangle$ | Eq. (4.29) |
| $\overleftrightarrow{\epsilon}_{kk'}$ | energy tensor for single particle non-condensate states | Eq. (4.27) |
| $\zeta_{CD}^{AB}$ | overlap integral of single particle wave functions $\Psi_A, \Psi_B, \Psi_C$ and $\Psi_D$ | Eq. (4.11) |
| $\eta_k$ | unperturbed single particle energies of $|\chi_k\rangle$ | Eq. (1.15) |
| $|\Theta_k\rangle$ | complete orthonormal non-condensate single particle basis | Eq. (4.22) |
| $\Lambda_{\rightsquigarrow}^{(\pm)}(N-N_0, T)$ | complex valued transition rate for single particle exchange events ($\rightsquigarrow$) | Eq. (5.40) |
| $\Lambda_{\leftrightsquigarrow}^{(\pm)}(N-N_0, T)$ | complex valued transition rate for pair exchange events ($\leftrightsquigarrow$) | Eq. (5.51) |

# NOTATION GUIDE

## Greek Letters

| | | |
|---|---|---|
| $\Lambda_{\circlearrowleft}^{(\pm)}(N-N_0,T)$ | complex valued transition rate for scattering events ($\circlearrowleft$) | Eq. (5.61) |
| $\lambda_{\leadsto}^{(\pm)}(N-N_0,T)$ | real valued transition rate for single particle exchange processes ($\leadsto$) | Eq. (5.41) |
| $\lambda_{\leadsto\leadsto}^{(\pm)}(N-N_0,T)$ | real valued transition rate for pair exchanges processes ($\leadsto\leadsto$) | Eq. (5.52) |
| $\lambda_{\circlearrowleft}^{(\pm)}(N-N_0,T)$ | real valued transition rate for scattering processes ($\circlearrowleft$) | Eq. (5.62) |
| $\lambda(T)$ | thermal de Broglie wave length | Eq. (1.4) |
| $\mu_0$ | eigenvalue of the Gross-Pitaevskii equation for $N$ particles | Eq. (4.4) |
| $\mu_\perp(N-N_0)$ | non-condensate chemical potential for $(N-N_0)$ particles at temperature $T$ | Eq. (7.18) |
| $\xi = a\varrho^{1/3}$ | perturbation parameter of the theory | Eq. (3.12) |
| $\varrho$ | atomic gas density | Eq. (1.3) |
| $\varrho_0$ | atomic condensate density | Eq. (7.49) |
| $\varrho_\perp$ | atomic non-condensate density | Eq. (7.49) |
| $\hat{\rho}_{GC}(\mu,T)$ | thermal state of the grand canonical ensemble | Eq. (1.16) |
| $\hat{\rho}_C(T)$ | thermal state of the canonical ensemble | Eq. (5.16) |
| $\rho_1(t)$ | single particle density matrix | Eq. (1.28) |
| $\hat{\rho}_0^{(N)}(t)$ | reduced condensate density matrix | Eq. (5.16) |
| $\hat{\rho}_\perp^{(N)}(t)$ | reduced non-condensate density matrix | Eq. (5.12) |
| $\hat{\sigma}^{(N)}(t)$ | $N$-body density matrix | Eq. (3.8) |
| $\sigma_0(t) = \langle N_0 \rangle(t)/N$ | condensate fraction | Eq. (6.4) |
| $\sigma_\perp(t) = \langle N_\perp \rangle(t)/N$ | non-condensate fraction | Eq. (6.4) |
| $\tau_0$ | time scale of condensate evolution | Eq. (3.1) |
| $\tau_{col}$ | average time scale for two body collisions | Eq. (3.1) |
| $|\Phi_k\rangle$ | eigenbasis of single particle density matrix | Eq. (1.28) |
| $|\chi_k\rangle$ | single particle eigenbasis of the non-interacting gas | Eq. (1.14) |
| $|\Psi_0\rangle$ | Gross-Pitaevskii wave function | Eq. (4.4) |
| $|\Psi_k\rangle$ | wave functions of non-condensate particles | Eq. (4.22) |
| $\hat{\Psi}$ | second quantized bosonic field | Eq. (4.5) |
| $\hat{\Psi}_0$ | second quantized bosonic condensate field | Eq. (4.5) |
| $\hat{\Psi}_\perp$ | second quantized bosonic non-condensate field | Eq. (4.5) |
| $\Psi_N(\mathbf{r}_1,\ldots,\mathbf{r}_N,t)$ | $N$-body wave function | Eq. (2.12) |

18

Chapter 1

# Bose-Einstein condensation in ideal Bose gases

We recall the technical terms "Bose-Einstein condensation", and "quantum ergodicity", before the reader is introduced into the experimental state-of-the-art. Einstein's original prediction of Bose-Einstein condensation is summarized in a short fashion to demonstrate the link of the Einstein de Broglie condition[1] to the first experimental observations of condensate formation [23, 24, 25]. Thereupon, the canonical and the grand canonical statistical ensembles are implemented as state-of-the-art theoretical techniques to access the condensate particle number statistics of non-interacting bosonic gases below the critical temperature $T_c$ for Bose-Einstein condensation.

## 1.1 What is a Bose-Einstein condensate?

*Encyclopic definition*: "When a gas of bosonic particles is cooled below a critical temperature $T_c$, it condenses into a Bose-Einstein condensate. The condensate

---
[1]The Einstein de Broglie condition results from the definition of a critical temperature in the original theory of Bose-Einstein condensation (see Section 1.3), and means that the average distance of the particles in the gas must be smaller than their de Broglie wavelength in order to observe condensate formation.

consists of a macroscopic number of particles, which are all in the ground state of the system. Bose-Einstein condensation (BEC) is a phase transition, which does not depend on the specific interactions between particles. It is based on the indistinguishability and wave nature of particles, both of which are at the heart of quantum mechanics [26]."

We shall recall here that the purpose of the present thesis is to directly model the microscopic condensate number distribution during the Bose-Einstein phase transition under inclusion of both the wave nature and the indistinguishability of the quantum particles. Within our theory, we will theoretically proof that the equilibrium steady state indeed depends on the specific (nonlinear) form of the interactions, nevertheless recovering the statistics of a thermal state for an ideal gas!

## 1.2 What is quantum ergodicity?

The expression "ergodicity" refers to a concept of classical statistical mechanics. Introduced by Ludwig Boltzmann in the nineteenth century [13, 14], a system which behaves ergodically is ment to sample each point in phase space equally over time, so that each state with the same energy has equal probability to be populated. Boltzmann showed that his conjecture applies for a gas of non-interacting, classical particles, subject to the condition of fixed energy and fixed particle number, evolving to a maximum entropy thermal state under the assumption of molecular chaos. However, some examples from classical statistical mechanics are known to be non-ergodic (e.g. strictly integrable systems) and do not relax into a thermal state, even after infinitely long times, such as a chain of coupled, one-dimensional harmonic oscillators [11]. Even less is known about the accuracy of the thermal state ansatz for quantum systems with finite particle number (where the density matrix does not necessarily factorize into different partitions such as condensate and non-condensate), especially for weakly interacting, quantum degenerate bosonic gases. So far, the ergodicity conjecture has been proven [27] only for ideal quantum gases coupled to an external heat reservoir. For an ideal gas, it is intuitive that the steady state of the non-interacting particles being in contact with a heat reservoir is a thermal state – independent of the condensate non-condensate interaction strength – since entirely the coupling to the external heat reservoir (which itself is in a thermal state) thermalizes the system. In contrast, the equilibrium steady state of a weakly interacting Bose gas which undergoes condensation because of atomic collisions as predicted by our master equation theory still depends on the specific nonlinearity of the atomic interactions: A question to be answered in the present thesis is hence whether a weakly interacting gas of finite particle number below $T_c$ really relaxes towards a thermal Boltzmann state of an ideal quantum gas [27], in the limit of very weak interactions, as presumed by the theory of thermodynamics?

## 1.3 Original prediction of Bose-Einstein condensation

In the 20's of the twentieth century, Einstein predicted [28, 29] what we call today "Bose-Einstein condensation": a macroscopic number expectation value of a single particle quantum state, in a gas of $N$ indistinguishable, non-interacting bosonic particles.

The heart of Bose's contribution [30, 31] to Bose-Einstein condensation was to treat a photon gas as an ensemble of indistinguishable bosonic particles, inspiring Einstein to apply [28, 29] Bose's statistics [30, 31] equivalently to ideal monoatomic gases enclosed in a volume $V$. This led him to the Bose-Einstein distribution function

$$N_{\vec{l}} = \frac{1}{\exp[\alpha + \beta \eta_{\vec{l}}] - 1} \ . \tag{1.1}$$

Equation (1.1) refers to the average occupation number $N_{\vec{l}}$ of a single particle state with energy $\eta_{\vec{l}} = \hbar^2 |\vec{l}|^2 / 2m$, where $\vec{l} = (l_x, l_y, l_z)$ is a particle's wave vector in each spatial direction x, y, and z, $\beta = (k_B T)^{-1}$ the inverse thermal energy of the gas, and $\alpha$ a Lagrangian multiplier. For a gas at thermal equilibrium, $\alpha$ can be interpreted [10] as the product of the inverse thermal energy $\beta$ and the chemical potential $\mu$ of the gas, defined by

$$\mu = -\beta^{-1} \frac{\partial \ln \mathscr{Z}(N, T)}{\partial N} . \qquad (1.2)$$

In Eq. (1.2), $\mathscr{Z}(N,T)$ denotes the partition function of $N$ indistinguishable, non-interacting bosonic atoms at temperature $T$, i.e. the number of different available microstates to the system, see Eqs. (1.7, 1.9). In thermodynamic terms, $\mu$ is the change of the Helmholtz free energy $\mathscr{F} = -\beta^{-1} \ln \mathscr{Z}(N,T)$ with the particle number, being proportional to the change in Boltzmann's entropy $\mathscr{S} = k_B \ln \mathscr{Z}(N,T)$.

Einstein speculated that the equilibrium state of a Bose gas — which is the state of maximum entropy and minimum free energy according to the postulates of thermodynamics [10] — reveals that all particles in the gas "condense" into the same quantum state, if the number of particles in the gas tends to infinity. Indeed, in the limit $N \to \infty$ at fixed temperature, we notice that the number of available microstates $\mathscr{Z}(N,T)$ in the gas does (intuitively) no longer change significantly with the particle number, so that the chemical potential in Eq. (1.2) approaches the single particle ground state energy of the gas, being zero for a non-interacting gas in a box.

According to Eq. (1.1), Einstein recognized that macroscopic average ground state occupation should especially occur for high particle densities[2] at fixed temperature. This can be retraced by imposing that the number of particles in the gas be constant, and by summing Eq. (1.1) over all possible values of $\vec{l}$ except the condensate single particle mode, $\vec{l} = (0,0,0) \equiv 0$. Replacing the summation by an integration over the density of states $g(\eta) = V m^{3/2}/2^{1/2} \pi^2 \hbar^3 \eta^{1/2}$ (see Section 7.3) and taking the limit $\mu \to 0^-$ (reflecting the behavior of $\mu$ in Eq. (1.2) in the limit $N \to \infty$),

---
[2]E.g. achieved by lowering the volume at fixed particle number, or by adding particles at constant volume

## 1.3. ORIGINAL PREDICTION OF BOSE-EINSTEIN CONDENSATION

the ground state occupation number in Eq. (1.1) diverges, if we match the Einstein de Broglie relation:

$$\varrho\lambda^3(T_c) = \zeta(3/2) = 2.612 \ . \tag{1.3}$$

Equation (1.3) arises from the requirement that the integral over all non-condensate single particle occupations in Eq. (1.1) equals the total number of particles, $N$, at the critical point of the phase transition. Here, $\zeta(y) = \sum_{k=1}^{\infty} k^{-y}$ is the Riemann Zeta function, see Table 7.1, $\varrho = N/V$ the (homogeneous) atomic density of the gas, and $\lambda(T)$ is the de Broglie wavelength of the particles:

$$\lambda(T) = \left( \frac{2\pi\hbar^2}{mk_B T} \right)^{1/2} . \tag{1.4}$$

Equation (1.3) indicates in particular that Bose-Einstein condensation occurs, if the wavelength $\lambda(T)$ of the quantum particles in the gas becomes larger than their mean inter particle distance.

By default, this condition is interpreted as the wave length of the atoms in the gas getting infinitely large such that all particles are supposed to overlap and to form a giant matter wave, the condensate. The first monitoring of the microscopic quantum dynamics in this thesis (see part III) reflects that the reaching of the Einstein de Broglie condition leads to fulminating non-condensate number fluctuations and an average macroscopic ground state occupation. Our microscopic, many particle picture thus partially reproduces the idealized, intuitive picture of the condensate to consist of one giant matter wave, however, reflecting the actual balancing process of particle flow towards and out of the condensate mode, garnished by large quantum fluctuations characteristic for the Bose-Einstein phase transition.

Note that the Bose-Einstein phase transition is in particular defined in the thermodynamic limit, $N \to \infty, V \to \infty$, with $\varrho = $ const., meaning that the particle number and the quantization volume simultaneously tend to infinity, such as to keep the atomic density $\varrho$ and the critical temperature $T_c$ fixed. In this limit, the result obtained in Eq. (1.3) becomes exact (recompensating the approximation for the density of states $g(\eta)$ to be continuous, see Chapter 8), defining analytically the transition temperature $T_c$ for Bose-Einstein condensation in a uniform,[3] non-interacting Bose gas:

$$T_c = \frac{2\pi\hbar^2 \varrho^{2/3}}{k_B \zeta(3/2)^{2/3} m} \,. \tag{1.5}$$

How was Einstein led to Eq. (1.1)? Having a look to the original predictions of Bose-Einstein condensation [28, 29], we recognize that the major underlying assumption is the indistinguishability of particles: The number of quantum cells (in phase space) with energies between $\eta_{i^*}$ and $\eta_{i^*} + \Delta\eta$ is

$$z_{i^*} = \frac{2\pi V}{h^3} (2m)^{3/2} \eta_{i^*}^{1/2} \Delta\eta \,. \tag{1.6}$$

According to Bose's previous analysis, Einstein infered [28] that the number of possibilities to distribute $N_{i^*}$ indistinguishable particles over $z_{i^*}$ cells within the infinitesimal energy interval $\Delta\eta$ is given by

$$\mathscr{L}_{i^*} = \frac{(N_{i^*} + z_{i^*} - 1)!}{N_{i^*}!(z_{i^*} - 1)!} \,. \tag{1.7}$$

This can be understood as follows [27]: Consider $N_{i^*}$ particles (drawn as a one-dimensional sequence of dots), and $z_{i^*}$ lines which represent the different cells (as vertical lines creating a certain partition of the one-dimensional row). The number of positions carrying a label in this one-dimensional row is $N_{i^*} + z_{i^*} - 1$, so that the number of different configurations having $N_{i^*}$ dots in $N_{i^*} + z_{i^*} - 1$ labels equals the number of *different* microstates, which is exactly the binomial coefficient in

---

[3] uniform ≡ non-interating gas in a box of volume $V$

## 1.3. ORIGINAL PREDICTION OF BOSE-EINSTEIN CONDENSATION

Eq. (1.7).

Taking into account all different energies $\eta_{\vec{i}}$, the total number of microstates is $\mathscr{Z}(N,T) = \prod_{\vec{i}} \mathscr{Z}_{\vec{i}}$, assuming that the state of the gas factorizes. Then, Einstein adopts the definition [10] of Boltzmann's entropy, $\mathscr{S} = k_B \ln \mathscr{Z}(N,T)$, where $k_B$ is the Boltzmann constant, which (with the above partition function) leads to the entropy [10]

$$\mathscr{S} = k_B \sum_{\vec{i}} \left[ N_{\vec{i}} \ln\left(1 + \frac{z_{\vec{i}}}{N_{\vec{i}}}\right) + z_{\vec{i}} \ln\left(\frac{N_{\vec{i}}}{z_{\vec{i}}} + 1\right) \right]. \tag{1.8}$$

Equation (1.1) is subsequently derived from maximizing $\mathscr{S}$ (by setting the first order variation of $\mathscr{S}$ to zero), under the constraint that $\sum_{\vec{i}} N_{\vec{i}} = N$ and $\sum_{\vec{i}} N_{\vec{i}} \eta_{\vec{i}} = E$. Hence, Einstein derived Eq. (1.1) by assuming a unique maximum entropy equilibrium state which can be factorized, treating the particles in the gas as indistinguishable, and neglecting number and energy fluctuations.

What does hence happen, Einstein asked, if the particles are considered as distinguishable? In that case, the number of possibilities to distribute $N_{\vec{i}}$ on $z_{\vec{i}}$ cells is simply

$$\mathscr{Z}_{\vec{i}} = (z_{\vec{i}})^{N_{\vec{i}}}, \tag{1.9}$$

that means, each of the $N_{\vec{i}}$ particles has the same probability of occupying any cell $z_{\vec{i}}$, irrespectively of a single particle state's occupation with energy $\eta_{\vec{i}}$, and $\vec{l} \neq \vec{k}$. Again, taking into account all energies as in Eq. (1.8), care has to be taken that a microstate with $\{N_{\vec{i}_1}, N_{\vec{i}_2}, \ldots\}$ particles occupying the cells $\{z_{\vec{i}_1}, z_{\vec{i}_2}, \ldots\}$ can be realized in $N!/\prod_{\vec{i}} N_{\vec{i}}!$ different ways, considering for a moment the particles as distinguishable. Hence, the total number of states is given by $\mathscr{Z}(N,T) = \prod_{\vec{i}} \mathscr{Z}_{\vec{i}} = N! \prod_{\vec{i}} (z_{\vec{i}})^{N_{\vec{i}}}/N_{\vec{i}}!$, which yields the Boltzmann entropy

$$\mathscr{S} = k_B \left[ N \ln N + \sum_{\vec{i}} \left[ N_{\vec{i}} \ln\left(\frac{z_{\vec{i}}}{N_{\vec{i}}}\right) + N_{\vec{i}} \right] \right] \tag{1.10}$$

by taking the natural logarithm. Equation (1.10) indicates that the resulting entropy cannot be correct, i.e., the number of possible microstates is overcounted. This is because the first term in Eq. (1.10) is proportional to $N \ln N$ — contradicting the extensivity property [10] of the thermodynamic entropy, $\mathscr{S}(\lambda N_1 + \mu N_2) = \lambda \mathscr{S}(N_1) + \mu \mathscr{S}(N_2)$. Moreover, modeling the limit of zero temperature by setting $N_0 \to N$, and $N_{\vec{l}} \to 0^+$, for all $\vec{l} \neq (0,0,0)$, the expression in Eq. (1.8) for indistinguishable particles gives the correct limit $\mathscr{S} \to 0^+$ (as imposed by the 3$^{rd}$ law of thermodynamics [10]), whereas Eq. (1.10) for distinguishable particles leads to $k_B N \ln N$.

The main assertion of Bose and Einstein in a nutshell was thus that radiation can be treated as a photon gas, with the same specific combinatoric results induced by indistinguishability.

## 1.4 Experimental state-of-the-art

As reported in Section 1.3, Einstein's original prediction refered to a gas of non-interacting particles in the thermodynamic limit $N \to \infty, V \to \infty$, with $\varrho = $ const. Thus, his prediction could not be taken for granted to work also for finite, interacting Bose gases in harmonic, typically anisotropic traps.[4] The solidification of almost all materials at typical densities required at usual (e.g. room) temperatures for the reaching of Einstein's condition in Eq. (1.3) is the major problem of realizing Bose-Einstein condensation experimentally [33].

To achieve Bose-Einstein condensation in the laboratory, the atomic ensemble is therefore brought to extremly low atomic densities by laser cooling [34] and is rapidly cooled hereupon to very low temperatures by evaporative cooling techniques [35, 36]. By this means, the gas has no time to solidify, whereas Einstein's condition in Eq. (1.3) can still be matched. Typical densities and temperature ranges required to achieve Bose-Einstein condensation are [15, 23, 24, 32]:

$$\varrho \sim 10^{12} - 10^{15} \text{cm}^{-3} \quad \text{and} \quad T \sim 20 \text{ nK} - 1 \text{ }\mu\text{K} \ . \tag{1.11}$$

---
[4]See Section 1.5 for the quantum statistics of non-interacting gases in harmonic traps.

## 1.4. EXPERIMENTAL STATE-OF-THE-ART

First observations of Bose-Einstein condensation in the laboratory were reported in 1995, for the alkali species $^{87}$Rb [23] in the group of Eric Cornell and Carl Wieman, at the Joint Institute for Laboratory Astrophysics [23], for $^{23}$Na [24] in the group of Wolfgang Ketterle, at the Massachusetts Institute of Technologies [32], and for $^{7}$Li [37] at RICE university. Up to date, Bose-Einstein condensation has been experimentally proven to exist in $^{1}$H, $^{7}$Li, $^{23}$Na, $^{39}$K, $^{52}$Cr, $^{85}$Rb, $^{133}$Cs, $^{170}$Yb and $^{4}$He [15].

Except for the species $^{4}$He [38, 39], which obeys – contrarily to all other summarized candidates – very strong interactions between its atomic constituents in the Bose condensed phase, the typical atomic density of a Bose-Einstein condensate is surprisingly dilute: At the center of the trap, where the highest atomic density (the condensate) is located, it is of the order of $\varrho \sim 10^{12} - 10^{15}$ cm$^{-3}$. In comparison, the density of air molecules at room temperature and atmospheric pressure is about four to seven orders of magnitudes larger [15]. A direct quantitative measure for the diluteness of a Bose gas is the gas parameter $\xi = a\varrho^{1/3}$ (where $a$ is the s-wave scattering length, see Section 2.1), typically of the order

$$\xi = a\varrho^{1/3} \sim 10^{-2} \ll 1 , \qquad (1.12)$$

for a dilute Bose-Einstein condensate. Thus, the experimental path of producing Bose-Einstein condensates becomes theoretically noticeable as a small parameter in our master equation ansatz in Part II of the thesis: The dilute gas parameter $\xi = a\varrho^{1/3}$ will be identified in the derived transition rates for particle exchange between the non-condensate and the condensate, and is employed to quantify condensate formation times in a perturbative approach for the condensate wave function. The same applies for the condensate and non-condensate steady state number distributions.

In the remainder of the thesis, state-of-the-art experimental parameters such as those of the early experiments on Bose-Einstein condensation [23, 24, 32] are used for quantitative calculations of condensate formation times and particle number

| physical parameter | JILA [$^{87}$Rb] | MIT [$^{23}$Na] | SI unit |
|---|---|---|---|
| atomic density $\varrho$ | $2.6 \times 10^{12}$ | $1.0 \times 10^{14}$ | cm$^{-3}$ |
| s-wave scattering length $a$ | $5.7 \times 10^{-9}$ | $4.9 \times 10^{-9}$ | m |
| gas parameter $\varrho a^3$ | $5.0 \times 10^{-7}$ | $1.2 \times 10^{-5}$ | |
| trap frequencies $\nu_x, \nu_y, \nu_z$ | 42.0, 42.0, 120.0 | 235.0, 410.0, 745.0 | s$^{-1}$ |
| total particle number $N$ | 2000 | $5 \cdot 10^5$ | |
| critical temperature $T_c$ | $\sim 32$ | $\sim 2000$ | nK |
| typical formation time $\tau_0$ | $\sim 2.0 - 4.0$ | $\sim 0.5 - 1.0$ | sec |

Table 1.1: *Typical parameters of the early experiments at JILA [23] and MIT [32], used for numerical calculations throughout the thesis. The meaning of the s-wave scattering length as given in the table is explained in Section 2.1.*

distributions during and after condensate formation. A recollection of relevant experimental parameters is shown in Table 1.1.

## 1.5 Bose-Einstein condensation in harmonic traps

In order to describe the statistics of a bosonic gas in an external confinement, the original analysis of Bose-Einstein statistics for uniform gases needs to be extended to harmonic traps. This is realized within the quantum version of the canonical and the grand canonical ensemble, which are conventually used to describe the statistics of non-interacting bosonic gases [10].

In classical thermodynamics, the two ensembles are equivalent in the thermodynamic limit of large particle numbers. Note, however, that an unsolved problem in the theory of quantum degenerate gases below the critical temperature is that the canonical and the grand canonical ensemble lead to different predictions for the condensate statistics, even in the thermodynamic limit [27]. Therefore, the results on condensate statistics below $T_c$ obtained from the grand canonical and the canonical ensemble shall be contrasted: Although both ensembles predict the same expectation value of the condensate particle number in the thermodynamic limit (and similar occupation for finite particle numbers), the grand canonical ensem-

## 1.5. BOSE-EINSTEIN CONDENSATION IN HARMONIC TRAPS

ble features the so called "fluctuation catastrophe" (divergence of the condensate particle number variance in the thermodynamic limit) below $T_c$. Hence, it is the canonical ensemble which is in accordance with experimentally observed scenarios for condensate particle number expectation values and condensate number variances below the critical temperature.[5]

### 1.5.1 Grand canonical ensemble

We consider a gas of non-interacting atoms in a harmonic trapping potential, described by the first quantized Hamiltonian

$$h_1(\vec{r}) = \frac{1}{2m}\left(p_x^2 + p_y^2 + p_z^2\right) + \frac{1}{2}m(\omega_x^2 x^2 + \omega_y^2 y^2 + \omega_z^2 z^2) - \frac{1}{2}(\hbar\omega_x + \hbar\omega_y + \hbar\omega_z) , \quad (1.13)$$

with trapping frequencies $\vec{\omega} = (\omega_x, \omega_y, \omega_z)$, momenta $\vec{p} = (p_x, p_y, p_z)$ of the atoms in the three different spatial directions $\vec{r} = (x, y, z)$ of Euklidian space $\mathbb{R}^3$. In Eq. (1.13), the zero point energy is substracted for convenience. The eigenvectors vectors $|\chi_{\vec{l}}\rangle$ of $h_1(\vec{r})$ are labeled by the three component vector $\vec{l} = (l_x, l_y, l_z)$, with $l_i \in \mathbb{N}_0$. For non-interacting systems, the single particle eigenstates $\langle \vec{r} | \chi_{\vec{l}} \rangle$ in position representation are given by

---

[5]This indicates that a physical description of the Bose gas should keep the number of particles fixed. This is due to the separation of time scales in the Bose-Einstein condensate, leaving classical number correlations of condensate and non-condensate because of particle number conservation: Since the thermalization dynamics in the non-condensate is much faster than condensate formation, the calculation of any observable $\langle \hat{X} \rangle$ for fluctuating total particle numbers should consist in calculating its average first for a fixed particle number $N$, taking the ensemble average of the $N - N_0$ non-condensate thermal particles for each state of $N_0 = 0 \ldots N$ condensate particles. Once the expectation value of the observable $\langle \hat{X} \rangle_N$ for a fixed $N$ is known, the average of $\langle \hat{X} \rangle_N$ over ensembles refering to different total particle numbers $N$ is to be carried out. Not least for this purpose, we keep the number of atoms in the Bose gas fixed to $N$ for deriving the equilibrium steady state of a Bose-Einstein condensate in Part II of the thesis.

$$\langle \vec{r}|\chi_{\vec{l}}\rangle = \prod_{\xi=x,y,z} \frac{1}{\sqrt{2^{l_\xi} l_\xi!}} \left(\frac{L_\xi^2}{\pi}\right)^{1/4} e^{-\frac{L_\xi^2 \xi^2}{2}} H_{l_\xi}(L_\xi \xi) , \qquad (1.14)$$

where $L_\xi = \sqrt{m\omega_\xi/\hbar}$ is the width of the harmonic oscillator ground state, and the $H_{l_\xi}(L_\xi \xi)$ denote Hermite polynomials [40]. The corresponding single particle eigenenergies $\eta_{\vec{l}}$ read

$$\eta_{\vec{l}} = l_x \hbar \omega_x + l_y \hbar \omega_y + l_z \hbar \omega_z . \qquad (1.15)$$

Since the particles do not interact by assumption, particle exchange between atoms occupying the different single particle eigenmodes $|\chi_{\vec{l}}\rangle$ is a consequence of coupling the gas to an external heat reservoir. In addition to the energy exchange, the grand canonical ensemble assumes particle exchange with the external reservoir to account for fluctuations of the total number of particles as sketched in Fig. 1.1.

Assuming quantum ergodicity (equal occupation probability for all states with the same energy, see Section 1.2), and neglecting quantum mechanical number and energy fluctuations in the thermodynamic limit, the thermodynamical state [10] of the Bose gas at equilibrium is given by the thermal state

$$\hat{\sigma}_{GC}(\mu, T) = \frac{1}{\mathcal{Z}_{GC}(\mu, T)} \exp\left[-\beta \left(\hat{\mathcal{H}} - \mu \hat{N}\right)\right] , \qquad (1.16)$$

where $\hat{\mathcal{H}} = \sum_{\vec{l}=0}^{\infty} \eta_{\vec{l}} \hat{N}_{\vec{l}}$ is the second quantized Hamiltonian of a non-interacting gas, $\mu$ the corresponding chemical potential, i.e. the change of the gas' free energy with the total particle number, and $\hat{N}$ the number operator of atoms in the trap. Moreover,

$$\mathcal{Z}_{GC}(\mu, T) = \sum_{\vec{l}=0}^{\infty} \sum_{N_{\vec{l}}=0}^{\infty} \langle \{N_{\vec{l}}\}| \exp\left[-\beta \left(\hat{\mathcal{H}} - \mu \hat{N}\right)\right] |\{N_{\vec{l}}\}\rangle = \prod_{\vec{l}=0}^{\infty} \frac{1}{\exp[\beta(\eta_{\vec{l}} - \mu)] - 1} \qquad (1.17)$$

## 1.5. BOSE-EINSTEIN CONDENSATION IN HARMONIC TRAPS

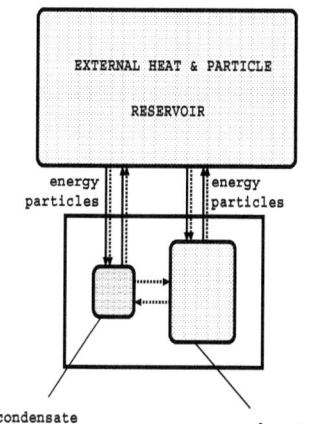

Figure 1.1: *Schematics of the grand canonical ensemble. Condensate and non-condensate are independently coupled to an external heat and particle reservoir. Particle flow between condensate and non-condensate is induced by equilibration of each subsystem (condensate and non-condensate) with the external reservoir. In the limit of vanishing atomic interactions, the thermodynamical steady state of maximum entropy under the constraint of fixed average energy and average particle number can still be reached, being a thermal state [10] independent of the interactions between the atoms, see Eq. (1.16).*

is the grand canonical partition function for indistinguishable particles,[6] accounting for normalization. Equation (1.17) is obtained by tracing $\exp\left[-\beta\left(\hat{\mathcal{H}} - \mu\hat{N}\right)\right]$ over all possible values of single particle occupations, $N_{\vec{i}} = 0 \ldots \infty$, $\forall \, \vec{i}$, and imposing particle number conservation onto the chemical potential $\mu$ [10].

The mean occupation number $\langle \hat{N}_{\vec{i}} \rangle$ of a single particle mode of energy $\eta_{\vec{i}}$ in the grand canonical ensemble is given by

$$\langle \hat{N}_{\vec{i}} \rangle = \frac{1}{\exp[\beta(\eta_{\vec{i}} - \mu)] - 1} = \prod_{j=x,y,z} \left[\frac{1}{z}\exp[\beta\hbar\omega_j] - 1\right]^{-1}, \quad (1.18)$$

---

[6]Distinguishable particles would imply a factor of $N!/\prod_{\vec{k}} N_{\vec{k}}!$ in each summand in Eq. (1.17), as explained in the derivation of Eq. (1.10).

where the fugacity $z = \exp[\beta\mu]$ is introduced with a range of variation $0 < z < 1$ (according to the chemical potential $\mu$ in Eq. (1.2), ranging from $\mu = -\infty$ to 0). The fugacity is a measure of the quantum degeneracy in the Bose gas: The classical limit (low concentration, i.e., low particle numbers and high temperatures, meaning that $\ln\mathscr{Z}(N,T)$ changes rapidly with $N$) exhibiting a large number of different possible states available to the system is formally accounted for by the limit $z \to 0^+$, meaning that $\mu \to -\infty$ according to the definition of the chemical potential $\mu$ in Eq. (1.2). Here, Boltzmann occupation numbers $\langle N_{\vec{i}}\rangle = \exp[-\beta\eta_{\vec{i}}]$ in Eq. (1.1) are recovered.

The quantum degenerate regime (low temperatures and high particle numbers), where the number of states changes only slightly with the particle number is reflected by the limit $\mu \to 0^-$. This implies $z \to 1^-$ and thus predicts Bose-Einstein condensation, i.e. a divergence of the average ground state occupation number, $\langle N_0\rangle \to \infty$. To evaluate the ground state occupation analytically in the quantum degenerate limit, first all average occupation numbers of excited (non-condensate) single particle states are counted,

$$\langle N_\perp\rangle = \sum_{\vec{i}\neq 0}\langle N_{\vec{i}}\rangle \simeq \left(\frac{k_B T}{\hbar\bar{\omega}}\right)^3 \zeta(3) \,, \tag{1.19}$$

with $\bar{\omega} = (\omega_x, \omega_y, \omega_z)^{1/3}$ the averaged trap frequency. To derive the right hand side of Eq. (1.19), the sum $\sum_{\vec{i}}$ is replaced by an integral $\int d\eta g(\eta)$, given the density of states $g(\eta) = \eta^2 2^{-1}(\hbar^3\omega_x\omega_y\omega_z)^{1/3}$ [15] for a three-dimensional harmonic trap. This ansatz for the density of states is strictly valid only for large particle numbers, where the approximation of the non-condensate single particle spectrum being quasi-continuous is recompensated by assuming a very large Bose gas ($N \sim 10^{23}$) thus formally using the thermodynamic limit (see Chapter 8).

Imposing particle number conservation (after the calculation), $\langle N_0\rangle + \langle N_\perp\rangle = N$, the result in Eq. (1.19) is rewritten in order to find the ground state occupation as

## 1.5. BOSE-EINSTEIN CONDENSATION IN HARMONIC TRAPS

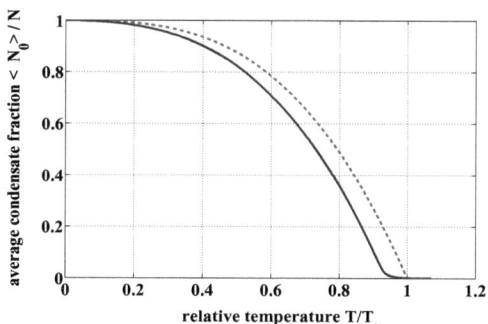

Figure 1.2: *Average condensate fraction $\langle N_0 \rangle /N$ predicted by the grand canonical result in Eq. (1.20) (red dashed line) vs. exact numerical calculations within the canonical ensemble using Eq. (1.27) (blue solid line). Calculations are performed for a gas of $N = 2500$ particles in a three-dimensional harmonic trap with trapping frequencies $\omega_x = \omega_y = 42.0$ Hz, $\omega_z = 120.0$ Hz. The ideal gas critical temperature $T_c = 36.47$ nK is defined by Eq. (1.3).*

a function of temperature in the grand canonical ensemble:

$$\frac{\langle N_0 \rangle}{N} = \left[ 1 - \left( \frac{T}{T_c} \right)^3 \right], \qquad (1.20)$$

with a critical temperature $T_c$ for a non-interacting Bose gas in a harmonic trap, given by

$$T_c = \frac{\hbar \bar{\omega} N^{1/3}}{k_B \zeta(3)^{1/3}}. \qquad (1.21)$$

The scaling behavior of the average condensate occupation number $\langle N_0 \rangle$ with $T/T_c$ in a three-dimensional harmonic trap differs from the scaling behavior for a homogenous gas in Eq. (1.5), i.e., the scaling is $(T/T_c)^3$ instead of $(T/T_c)^{3/2}$. This is due to the external confinement which induces higher condensate occupations, measuring the temperature in units of the ideal gas critical temperature $T_c$. Thus, the external trap confines the particles in the trap stronger, which leads to a larger condensate fraction as compared to the uniform case for the same gas temperature at equilibrium. In turn, a lower temperature is needed for the case of no external confinement in order to observe the same condensate fraction.

The average condensate occupation number $\langle N_0 \rangle$ versus $T/T_c$ predicted by the grand canonical ensemble result in Eq. (1.20) is illustrated in Fig. 1.2 (red dashed line), and compared with the exact calculation of Section 1.5.2 (canonical ensemble) in a harmonic trap using the condensate number distribution in Eq. (1.27). Whereas the grand canonical calculation of $\langle N_0 \rangle$ shows a cusp at the transition temperature $T_c$ of Bose-Einstein condensation, the canonical ensemble predicts condensate occupations only for temperatures below the ideal gas critical temperature $T_c$ (at $\sim 0.95 T_c$), and a smooth transition. These deviations originate from the replacement of the discrete sum by an integration to derive Eq. (1.20) under the assumption of a quasi-continuous spectrum. This results effectively in a shift of $T_c$, which is smaller than 5% starting at $N \sim 10000$, and ranges from 5 – 30% for smaller total particle numbers, starting from the percent level at $T = 0.2 T_c$ to approx. 20% at $T = 0.95 T_c$ in Fig. 1.2. This shift can be incorporated in Eq. (1.20) by replacing $T_c \rightarrow T_c \times (1 - 0.7275/N^{1/3})$, or by respecting the discreteness of the single particle spectrum via an exact numerical treatment as we do in Fig. (1.2), blue line (see also Chapter 8).

Albeit the average condensate occupation in Eq. (1.20) is correctly described in the grand canonical ensemble, it was soon recognized [41] that a grand canonical description of the gas cannot be correct below the critical temperature. This is because of the so called "grand canonical fluctuation catastrophe", which has been discussed by generations of physicists [42]. In short terms, the problem of the grand canonical ensemble below the critical temperature is that the variance of the condensate particle number, given analytically as

$$\Delta^2 N_0 = \langle N_0 \rangle (\langle N_0 \rangle + 1) \,, \tag{1.22}$$

## 1.5. BOSE-EINSTEIN CONDENSATION IN HARMONIC TRAPS

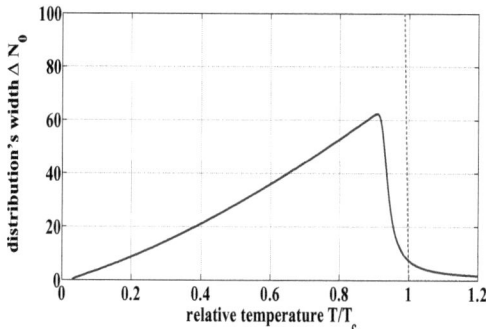

Figure 1.3: *Standard deviation $\Delta N_0 = (\langle N_0^2 \rangle - \langle N_0 \rangle^2)^{1/2}$ of the condensate number occupation, predicted by the grand canonical ensemble result in Eq. (1.22) (red dashed line) vs. exact numerical calculations within the canonical ensemble (blue solid line) using Eq. (1.25, 1.27), as a function of relative temperature $T/T_c$, for the same experimental parameters as in Fig. 1.2: The grand canonical ensemble predicts condensate number variances $\Delta^2 N_0$ as large as $N^2$ below $T_c$.*

where $\langle N_0 \rangle$ is given by Eq. (1.20), becomes comparable to the total particle number and therefore diverges in the limit $\langle N_0 \rangle \to N$. These large fluctuations ($\sqrt{\Delta^2 N_0} \sim N$) are contradictory to experimental observations, where the condensate number variance has been experimentally measured [43] to be in the Poisson to sub-Poisson range (hence $\sqrt{\Delta^2 N_0} \sim \sqrt{N}$).

Nowadays, the most reliable and numerically accessible state-of-the-art thermodynamic prediction for the condensate number variance is thus governed by the canonical ensemble for non-interacting gases below $T_c$, where the grand canonical and canonical ensemble cease to be equivalent [27]. The standard deviation of the condensate particle number $\sqrt{\Delta N_0^2}$ obtained within the grand canonical ensemble is shown in Fig. 1.3 as a function of $T/T_c$ (red dashed line), in comparison to our numerical prediction within the canonical ensemble discussed in the next section.

36 Chapter 1. BOSE-EINSTEIN CONDENSATION IN IDEAL BOSE GASES

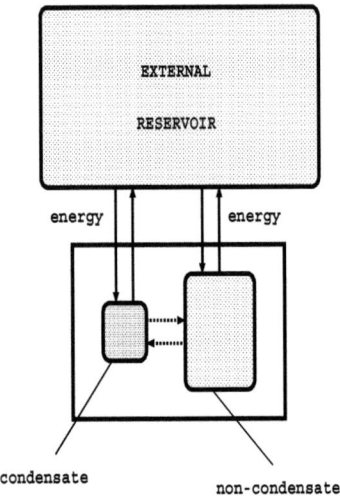

Figure 1.4: *Schematics of the canonical ensemble. Condensate and non-condensate are independently coupled to an external heat reservoir. Particle flow between condensate and non-condensate is induced by the energy exchange of either one subsystem (condensate and non-condensate) with the external heat reservoir. In the limit of vanishing interparticle interactions, the maximum entropy equilibrium state of the Bose gas can therefore still be reached, and is independent of the interacting strength, see Eq. (1.23).*

### 1.5.2 The canonical ensemble

The equilibrium state of the Bose gas in the canonical ensemble is derived under the constraint of a fixed total average energy $\langle E \rangle$ and a fixed *constant* particle number $N$ in the system, as sketched in Fig. 1.4. Within the ergodic assumption (see Section 1.2), and under the neglect of energy fluctuations in the thermodynamic limit, the (maximum entropy) equilibrium state of the gas is a thermal one [10, 27],

$$\hat{\sigma}_C^{(N)}(T) = \mathscr{D}_N \frac{\exp\left[-\beta\hat{\mathcal{H}}\right]}{\mathscr{Z}(N,T)} \mathscr{D}_N \, , \qquad (1.23)$$

## 1.5. BOSE-EINSTEIN CONDENSATION IN HARMONIC TRAPS

where $\hat{\mathcal{H}} = \sum_{\vec{k}} \eta_{\vec{k}} \hat{N}_{\vec{k}}$ is the many particle Hamiltonian of the ideal gas, $\hat{\mathcal{D}}_N$ a projector onto the Fock space of $N$ particles, and $\mathcal{Z}(N, T)$ the partition function of $N$ indistinguishable particles in the canonical ensemble:

$$\mathcal{Z}(N, T) = \text{Tr}\left\{\hat{\mathcal{D}}_N \exp\left[-\beta \hat{\mathcal{H}}\right] \hat{\mathcal{D}}_N\right\} = \sum_{\{N_{\vec{l}}\}}^{(N)} \langle\{N_{\vec{l}}\}|\exp\left[-\beta \hat{\mathcal{H}}\right]|\{N_{\vec{l}}\}\rangle . \quad (1.24)$$

The symbol $\sum_{\{N_{\vec{l}}\}}^{(N)}$ labels a partial sum over all tuples $\{N_{\vec{l}}, \vec{l} \in \mathbb{N}_0^3\}$ which satisfy $\sum_{\vec{l}=0}^{\infty} N_{\vec{l}} = N$. Clearly, this partition function differs from the standard ones for distinguishable particles by the missing prefactor $N!/\prod_{\vec{l}} N_{\vec{l}}!$. This factor needs to be included for distinguishable particles in order to realize that a Fock state $|\{N_{\vec{l}}\}\rangle$ has $N!/\prod_{\vec{l}} N_{\vec{l}}!$ different microscopic realizations, if we considered the particles as individuals. As we have seen in Section 1.3, however, this is not correct in the quantum degenerate limit, so that the partition function of indistinguishable particles in Eq. (1.26) has to be applied.

To access the condensate statistics, we deduce the condensate number distribution $p_N(N_0, T)$ from the diagonal element of the reduced density matrix in Eq. (1.23), where the trace is taken over all number states of the non-condensate which conserve the total number of particles:

$$p_N(N_0, T) = \langle N_0|\text{Tr}_\perp \hat{\sigma}_C^{(N)}|N_0\rangle = e^{-\beta \eta_0 N_0} \frac{\mathcal{Z}_\perp(N - N_0, T)}{\mathcal{Z}(N, T)} , \quad (1.25)$$

where $\mathcal{Z}_\perp(N - N_0, T)$ is the partition function of the non-condensate, containing $(N - N_0)$ particles:

$$\mathcal{Z}_\perp(N - N_0, T) = \sum_{\{N_{\vec{l}}\}, \vec{l} \neq 0}^{(N - N_0)} \langle\{N_{\vec{l}}\}|\exp\left[-\beta \hat{\mathcal{H}}_\perp\right]|\{N_{\vec{l}}\}\rangle , \quad (1.26)$$

with $\hat{\mathcal{H}}_\perp = \sum_{\vec{k} \neq 0} \eta_{\vec{k}} \hat{N}_{\vec{k}}$. Finally, using Eq. (1.25), we can derive an exact recurrence relation for $p_N(N_0, T)$:

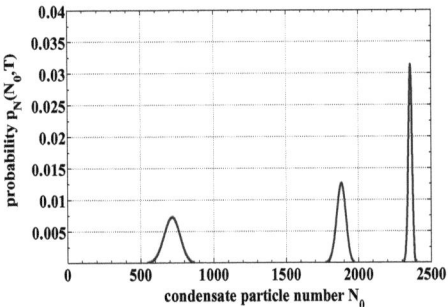

Figure 1.5: *Condensate particle number distribution within the canonical ensemble for a 3-dimensional harmonic trap with $\omega_x = \omega_y = 2\pi \times 42.0$ Hz, $\omega_z = 2\pi \times 120.0$ Hz, and $N = 2500$ $^{87}$Rb atoms, obtained from $p_N(N_0, T)$ in Eq. (1.27), for the same parameters as in Fig. 1.2, and for three different temperatures $T = 30.0, 20.0, 10.0$ nK (from left to right).*

$$\frac{p_N(N_0, T)}{p_N(N_0+1, T)} = e^{\beta \eta_0} \frac{\mathscr{Z}_\perp(N-N_0, T)}{\mathscr{Z}_\perp(N-N_0-1, T)} \ . \tag{1.27}$$

Some examples of the condensate particle number distribution $p_N(N_0, T)$ in Eq. (1.27) are shown in Fig. 1.5. As mentioned above, the average condensate occupations and the standard deviations of the condensate particle number as obtained from Eq. (1.27) are displayed in Figs. 1.2 and 1.3, respectively, comparing the grand canonical prediction (red dashed lines) to the canonical prediction (blue solid lines).

## 1.6 Bose-Einstein condensation in position space

In order to theoretically verify that Bose-Einstein condensation, i.e. a macroscopic condensate number expectation value of the single particle ground state mode can be related to the experimentally observed occurrence of a high condensate density

## 1.6. BOSE-EINSTEIN CONDENSATION IN POSITION SPACE

at the center of the trap below the critical temperature [23, 32], it is possible to apply the concept of the reduced one-body density operator as follows.

The bosonic gas of $N$ particles, whether interacting or not, is fully described by the many particle $N$-body state $\hat{\sigma}^{(N)}(t)$. The one-body density matrix, $\hat{\rho}_1$ [44], is the statistical operator of one particle in the quantum gas, averaged over all permutations of the $(N-1)$ remaining particles. It is defined by the quantum mechanical average

$$\hat{\rho}^{(1)}(t) = N \int_{\mathscr{C}} d^3\vec{r}_2 \ldots d^3\vec{r}_N \, \langle \vec{r}_2 \ldots \vec{r}_N | \hat{\sigma}^{(N)}(t) | \vec{r}_2 \ldots \vec{r}_N \rangle , \qquad (1.28)$$

where $\mathscr{C}$ denotes the volume of non-vanishing atomic density in the external trapping confinement. The diagonal elements of the one-body density matrix in position representation, $\langle \vec{r} | \hat{\rho}_1 | \vec{r} \rangle$, provide the average particle density in the trap. Off-diagonal elements, $\langle \vec{r} | \hat{\rho}_1 | \vec{r}\,' \rangle$, characterize spatial correlations in the Bose gas. The one-body density matrix is conveniently normalized to $N$.

We now take a closer look on ideal gases in a harmonic potential. In this case, the diagonal elements of the one-body density operator in spatial representation turn into

$$\langle \vec{r} | \hat{\rho}^{(1)} | \vec{r} \rangle = \sum_{\vec{i}} \langle N_{\vec{i}} \rangle \chi_{\vec{i}}^*(\vec{r}) \chi_{\vec{i}}(\vec{r}) , \qquad (1.29)$$

with the harmonic oscillator states $\chi_{\vec{i}}(\vec{r}) = \langle \vec{r} | \chi_{\vec{i}} \rangle$ in Eq. (1.14), and average single particle occupations $\langle N_{\vec{i}} \rangle$ given by Eq. (1.18). Assuming thermal equilibrium, we can write the one-body density matrix as

$$\hat{\rho}^{(1)} = \frac{1}{z^{-1}\exp(\beta h_1) - 1} , \qquad (1.30)$$

using the spectral decomposition of the first quantized Hamiltonian $h_1$ in Eq. (1.13), and the definition of the fugacity $z$ in Eq. (1.18).

To calculate the representation in position space $|\vec{r}\rangle$, we assume for simplicity an isotropic trapping potential, and use the expansion $\hat{\rho}^{(1)} = \sum_{m=1}^{\infty} z^m e^{-\beta m h_1}$ to arrive at [45]:

$$\langle \vec{r}|\hat{\rho}^{(1)}|\vec{r}\rangle = \left(\frac{m\omega}{\pi\hbar}\right)^{3/2} \sum_{k=1}^{\infty} \frac{z^k}{(1-e^{-2\beta k\hbar\omega})^{3/2}} \exp\left[-\frac{m\omega|\vec{r}|^2}{\hbar}\tanh\left(\frac{\beta k\hbar\omega}{2}\right)\right]. \qquad (1.31)$$

Equation (1.31) highlights the occurrence of the condensate part of the gas at the center of the trap, as the fugacity tends to unity below the transition temperature for Bose-Einstein condensation, $z \to 1^-$: Large summation indices $k$ in the sum entail a contribution close to one, i.e., the term $(1-e^{-2\beta k\hbar\omega})^{-3/2} \to 1$, and moreover, the term $\tanh(\beta k\hbar\omega/2) \to 1$. Around the center of the trap, $|\vec{r}| \to 0^+$, the sum in Eq. (1.31), and consequently the atomic density diverges in the limit $z \to 1^-$, whereas it tends to zero for distances larger than the harmonic oscillator length, $|\vec{r}| \gg \sqrt{\hbar/m\omega}$. Condensation onto the single particle ground state mode $\chi_0(\vec{r})$, i.e., an expectation value of $\langle N_0 \rangle \sim N$, hence manifests itself as enhanced atomic density at the center of the trap in harmonic trapping potentials.

It is evident that the situation is modified in the interacting case, which exhibits nevertheless a smooth transition [45] into the non-interacting case for sufficiently weak interactions ($a/LN \ll 1$, where $L$ is the extension of the harmonic oscillator ground state). In this case, single particle wave functions of the interacting system do not significantly differ from the harmonic oscillator states in Eq. (1.14).

Chapter 2

# Interacting Bose-Einstein condensates

In this Chapter, important concepts for treating interacting Bose-Einstein condensates are summarized. Under the restriction to two body interactions, justified for dilute atomic gases, the microscopic derivation of an effective interaction strength $g$ for atomic interactions in the Bose gas is sketched in Section 2.1. The Hamiltonian of an interacting Bose gas in second quantization is specified in Section 2.2. The Gross-Pitaevskii equation [15] constitutes a closed equation for the condensate mode in dilute atomic gases [46, 47, 48, 49]. It will be derived in Section 2.3 within the Hartree ansatz for the $N$-body state of the system. Section 2.4 finally summarizes theories for the study of the average condensate growth to establish the relation and contribution of the master equation theory of this thesis.

## 2.1 S-wave scattering approximation

In dilute atomic gases, it is possible to neglect three-body and higher order interactions [50], since atomic collisions are captured by successive two body interactions between the atoms in the Bose gas [16, 51]. These are described [52] by the first quantized Hamiltonian

$$h_{12} = \frac{\vec{p}^2}{2M} + V_{12}(\vec{r}) ,\qquad(2.1)$$

where $\vec{r} = \vec{r}_1 - \vec{r}_2$ is the relative coordinate, and $\vec{p} = \vec{p}_1 - \vec{p}_2$ is the relative momentum of the two identical colliding particles, while $M = m/2$ is their reduced mass, and $V_{12}(\vec{r})$ the two body interaction potential. From standard scattering theory [45, 52], we know that scattering states with positive energy $E = \hbar^2 k^2/2M$ obey the differential equation

$$\left(\Delta_{\vec{r}} + k^2\right)\Psi(\vec{r}) = \frac{2M}{\hbar^2} V_{12}(\vec{r})\Psi(\vec{r}) \,, \tag{2.2}$$

with the Laplacian operator $\Delta_{\vec{r}}$, and $k = |\vec{p}_1 - \vec{p}_2|/\hbar$, the absolute value of the relative wave vector between the two identical particles. Equation (2.2) has the formal solution:

$$\Psi(\vec{r}) = \Psi^{(in)}(\vec{r}) - \frac{2M}{4\pi\hbar^2} \int d\vec{r}\,' \frac{e^{ik|\vec{r}-\vec{r}\,'|}}{|\vec{r}-\vec{r}\,'|} V_{12}(\vec{r}\,')\Psi(\vec{r}\,') \,, \tag{2.3}$$

where $\Psi^{(in)}(\vec{r})$ is the unperturbed wave function of the incoming particle, satisfying the differential equation $\Delta_{\vec{r}}\Psi^{(in)}(\vec{r}) = -k_{in}^2 \Psi^{(in)}(\vec{r})$, with $k_{in}^2$ as the wave vector of the incoming wave ($|\vec{k}_{in}| = k_{in}$). For sufficiently weak interactions, we employ the first order (in the interaction $V_{12}$) Born approximation [52], which effectively consists in replacing $\Psi(\vec{r}\,') \to \Psi^{(in)}(\vec{r}\,')$ on the right hand side of Eq. (2.3). This approximation is valid for sufficiently weak interactions, as discussed below. There are two main ingredients used to derive an effective interaction strength for two body collisions:

*First*, assuming a sufficiently short, finite range interaction potential between the atoms, with an effective radius $R$, the asymptotic behavior of the scattering state for large distances $r \equiv |\vec{r}| \gg R$ between the two atoms turns into

$$\Psi(\vec{r}) = \Psi^{(in)}(\vec{r}) - \frac{e^{ikr}}{r} f(\vec{n}) \,, \tag{2.4}$$

where $\vec{n} = \vec{r}/r$ is the direction of scattering, and where

## 2.1. S-WAVE SCATTERING APPROXIMATION

$$f(\vec{n}) = -\frac{2M}{4\pi\hbar^2}\int_{\mathscr{C}} d\vec{r}\,' \, e^{ik\vec{n}\vec{r}\,'} V_{12}(\vec{r}\,')\Psi^{(in)}(\vec{r}\,') \qquad (2.5)$$

represents the scattering amplitude for the given scattering process. This amplitude does not depend on the relative distance $r$ between the atoms in the asympotic scattering region $r \gg R$.

*Second*, the limiting case of low energy collisions with a kinetic energy $\hbar^2 k^2/2M$ of the atoms of the order of the thermal energy $k_B T$ (with $T \leq 1.0\ldots 2.0\,\mu K$) is much smaller than the typical centrifugal barrier energy around $T = 1$ mK to scatter into higher angular momentum states than $l = 0$ [53]. In this case, the scattering amplitude does not depend on the direction $\vec{n}$ of scattering, and is set constant $f(\vec{n}) \to -a = \mathrm{const}$. The scattered part of the wave is thus rotationally symmetric, see Eq. (2.4), and $a$ is called "s-wave scattering length" as the strength of the interaction does not depend on the angle between the two scattered identical particles to lowest order, since only scattering states with angular momentum $l = 0$ are supposed.

Since ab-initio calculations of the s-wave scattering amplitude $a$ for realistic interatomic interaction potentials $V_{12}(\vec{r})$ are difficult tasks on their own [45, 53], we use a pseudo-potential [54],

$$V_{12}(\vec{r}_1 - \vec{r}_2) = g\delta(\vec{r}_1 - \vec{r}_2)\,, \qquad (2.6)$$

with $g = 4\pi\hbar^2 a/m$, in order to satisfy the same properties as the derived s-wave scattering amplitude $f(\vec{n}) = -a$, in the asymptotic limit of large distances $r \gg R$ and low kinetic energies $kR \ll 1$. The actual value of the s-wave scattering length $a$ is taken as an experimentally determined parameter, according to Table 1.1.

Using the s-wave scattering pseudo potential in Eq. (2.6) to describe two body collisions is justified [45], if $\xi = a\varrho^{1/3} \ll 1$, called "dilute gas condition". As noticed before, the dilute gas condition is satisfied in most state-of-the-art experiments for alkali atoms (see Section 1.4).

## 2.2 Hamiltonian for two body interactions

Given the effective description of two body interactions in Section 2.1 in terms of the s-wave scattering length, the formalism of second quantized bosonic fields can be introduced to describe the interacting Bose gas. Let's consider a gas of bosonic particles, each of which may be in a particular state of an orthonormal and complete set of single particle wave functions $\{|v_k\rangle, k \in \mathbb{N}_0\}$. Corresponding to the basis states $|v_k\rangle$, we introduce annihilation/creation operators, $\hat{c}_k$ and $\hat{c}_k^\dagger$, respectively, which satisfy bosonic commutation relations, $\left[\hat{c}_k, \hat{c}_l^\dagger\right] = \delta_{kl}$. The operators $\hat{c}_k$ and $\hat{c}_k^\dagger$ create particle states such as plane waves, or the harmonic oscillator states $|v_k\rangle \to |\chi_{\vec{k}}\rangle$ in Eq. (1.14), depending on the choice of the basis.

The operators $\hat{\Psi}(\vec{r})$ and $\hat{\Psi}^\dagger(\vec{r})$ are called "bosonic field operators" [55], which describe the quantized field of the gas, and satisfy the commutation relations

$$\left[\hat{\Psi}(\vec{r}), \hat{\Psi}(\vec{r}\,')\right] = \left[\hat{\Psi}^\dagger(\vec{r}), \hat{\Psi}^\dagger(\vec{r}\,')\right] = 0 \quad \text{and} \quad \left[\hat{\Psi}(\vec{r}), \hat{\Psi}^\dagger(\vec{r}\,')\right] = \delta(\vec{r} - \vec{r}\,') . \qquad (2.7)$$

The interpretation of the fields $\hat{\Psi}(\vec{r})$ and $\hat{\Psi}^\dagger(\vec{r})$ is that they annihilate and create, respectively, a bosonic particle at position $\vec{r}$. Expanding the state $|\vec{r}\rangle$ of a particle at position $\vec{r}$ in the orthonormal, complete basis $\{|v_k\rangle, k \in \mathbb{N}_0\}$, the ket $|\vec{r}\rangle = \sum_k |v_k\rangle\langle v_k|\vec{r}\rangle$ translates in particle number representation into

$$|\vec{r}\rangle = \hat{\Psi}^\dagger(\vec{r})|0\rangle = \sum_k \langle v_k|\vec{r}\rangle \, \hat{c}_k^\dagger|0\rangle . \qquad (2.8)$$

Hence, the two bosonic fields $\hat{\Psi}(\vec{r})$ and $\hat{\Psi}^\dagger(\vec{r})$ are defined[1] by

---

[1] Note that the second quantized field represents a particle and a wave simultaneously. The first can be associated to the particle operators $\hat{c}_k$ and $\hat{c}_k^\dagger$, whereas the wave nature

## 2.3. GROSS-PITAEVSKII EQUATION FROM THE HARTREE ANSATZ

$$\hat{\Psi}^\dagger(\vec{r}) = \sum_k v_k^*(\vec{r})\hat{c}_k^\dagger \quad \text{and} \quad \hat{\Psi}(\vec{r}) = \sum_k v_k(\vec{r})\hat{c}_k \ . \tag{2.9}$$

This definition entails the bosonic commutation relations for the fields $\hat{\Psi}(\vec{r})$ and $\hat{\Psi}^\dagger(\vec{r})$ in Eq. (2.7), given that the creation and annihilation operators $\hat{c}_k^\dagger$ and $\hat{c}_k$ satisfy bosonic commutation relations.

The Hamiltonian of a gas of bosonic particles, including two body interactions, is specified [55] in terms of the quantized fields $\hat{\Psi}(\vec{r})$ and $\hat{\Psi}^\dagger(\vec{r})$ as

$$\hat{\mathcal{H}} = \int_{\mathscr{C}} d\vec{r}\, \hat{\Psi}^\dagger(\vec{r}) \left[ \frac{-\hbar^2 \vec{\nabla}^2}{2m} + V_{\text{ext}}(\vec{r}) \right] \hat{\Psi}(\vec{r}) + \frac{g}{2} \int_{\mathscr{C}} d\vec{r}\, \hat{\Psi}^\dagger(\vec{r})\hat{\Psi}^\dagger(\vec{r})\hat{\Psi}(\vec{r})\hat{\Psi}(\vec{r}) \ . \tag{2.10}$$

In Eq. (2.10), such as in the sequel of this thesis, the effective two body s-wave scattering interaction potential $V_{12}(\vec{r}-\vec{r}\,') = g\delta(\vec{r}-\vec{r}\,')$ of Eq. (2.6) is adapted. The region of non-vanishing spatial atomic density is denoted by $\mathscr{C}$.

## 2.3 Gross-Pitaevskii equation from the Hartree ansatz

Here, the Gross-Pitaevskii equation [15, 16] is introduced, constituting a closed equation for the macroscopically occupied single particle mode $\Psi_0(\vec{r},t)$ below $T_c$, called the "condensate wave function". Assuming $N$ particles in the Bose gas to share the same, in general time dependent single particle mode $\Psi_0(\vec{r},t)$, the Hartree ansatz [15] can be used to derive the Gross-Pitaevskii equation. The first quantized $N$-particle Hamiltonian $\mathcal{H}_N$ for $N$ atoms interacting via two body collisions in the Bose gas is given by

---

is typified by the basis of wave functions $\{v_k(\vec{r})\}$, in which we expand the field $\hat{\Psi}^\dagger(\vec{r})$. The choice of this basis is in general arbitrary.

$$\mathcal{H}_N = \sum_{k=1}^{N}\left[\frac{\vec{p}_k^2}{2m}+V_{\text{ext}}(\vec{r}_k)\right]+\frac{g}{2}\sum_{k\neq l}\delta(\vec{r}_k-\vec{r}_l)\,, \qquad (2.11)$$

where $V_{\text{ext}}$ is the external trapping potential, and the factor 1/2 ensures that each pair of particles contributes only once (independent of the order of $k,l$) in Eq. (2.11). Assuming that all particles share the same quantum state at $T=0$, the Hartree-Fock ansatz is employed for the $N$-body ket of the Bose gas:

$$\Psi_N(\vec{r}_1,\ldots,\vec{r}_N,t) = \prod_{k=1}^{N}\Psi_0(\vec{r}_k,t)\,. \qquad (2.12)$$

Now, we can calculate the expectation value of the $N$-particle Hamiltonian in Eq. (2.11) with respect to the $N$-body ket $|\Psi_N\rangle$. According to Eqs. (2.11, 2.12), the latter is given as a functional $\mathcal{E}_N = \mathcal{E}_N(\Psi_0,\Psi_0^*)$ of the condensate wave function $\Psi_0$, and its conjugate $\Psi_0^*$:

$$\mathcal{E}_N(\Psi_0,\Psi_0^*) = N\int_{\mathscr{C}}d\vec{r}\left[\frac{\hbar^2}{2m}|\vec{\nabla}\Psi_0(\vec{r},t)|^2+V_{\text{ext}}(\vec{r})|\Psi_0(\vec{r},t)|^2+\frac{g(N-1)}{2}|\Psi_0(\vec{r},t)|^4\right]\,. \qquad (2.13)$$

In Eq. (2.13), the term $N(N-1)/2|\Psi_0(\vec{r},t)|^4$ describes the two body interactions between the particles, thus proportional to the number of $N(N-1)/2$ ways to pair the bosons, times the corresponding single particle densities $|\Psi_0(\vec{r},t)|^2$ of each boson contributing to a two body collision process. To derive an evolution equation for the condensate wave function $\Psi_0(\vec{r},t)$, one uses Hamiltonian's principle of least action [15, 56] (in analogy to the derivation of the Schrödinger equation), with the Lagrangian

$$\mathscr{L}(\Psi_0,\Psi_0^*) = \int_{\mathscr{C}}d\vec{r}\,\frac{i\hbar}{2}\left(\Psi_0^*\frac{\partial\Psi_0}{\partial t}-\Psi_0\frac{\partial\Psi_0^*}{\partial t}\right)-\mathcal{E}_N(\Psi_0,\Psi_0^*)\,. \qquad (2.14)$$

## 2.3. GROSS-PITAEVSKII EQUATION FROM THE HARTREE ANSATZ

Demanding the action $\int_{t_1}^{t_2} \mathscr{L} dt$ to get extremal, thus insisting that $\delta \int_{t_1}^{t_2} \mathscr{L} dt = 0$, and imposing vanishing variations $\delta\Psi_0$ and $\delta\Psi_0^*$ at the temporal boundaries $t = t_1$ and $t = t_2$ and at the spatial boundaries of the region $\mathscr{C}$ of non-vanishing gas density leads to the time dependent Gross-Pitaevskii equation

$$i\hbar \frac{\partial \Psi_0(\vec{r},t)}{\partial t} = \left[ \frac{-\hbar^2 \vec{\nabla}^2}{2m} + V(\vec{r}) + gN|\Psi_0(\vec{r},t)|^2 \right] \Psi_0(\vec{r},t) . \quad (2.15)$$

Equation (2.15) quantitatively determines the macroscopically occupied mode $\Psi_0(\vec{r},t)$ of the gas below $T_c$, termed the "condensate mode".

For a static condensate mode, with $N$ particles of the gas occupying this mode, the Gross-Pitaevskii wave function $\Psi_0(\vec{r},t)$ in Eq. (2.15) will evolve in time only with respect to a trivial phase factor, $\Psi_0(\vec{r},t) = \Psi_0(\vec{r}) e^{i\mu_0 t/\hbar}$, $\mu_0$ being defined as the eigenvalue of the static Gross-Pitaevskii equation:

$$\left[ \frac{-\hbar^2 \vec{\nabla}^2}{2m} + V_{\text{ext}}(\vec{r}) + gN|\Psi_0(\vec{r})|^2 - \mu_0 \right] \Psi_0(\vec{r}) = 0 . \quad (2.16)$$

In view of the quantum master equation theory of Bose-Einstein condensation in Part II of this thesis, we adapt one issue which arises from the above mean field theory:[2] The linear part of the Gross-Pitaevskii equation is identical to the well-known Schrödinger equation, whereas the nonlinear part reflects the presence of

---

[2]Mean field theories are often termed as such because they treat the dynamics of the Gross-Pitaevskii wave function $\Psi_0$ for a given average number of condensate particles, $N \to \langle N_0 \rangle$ in Eq. (2.15), representing the mean value of condensate particles which populate the average condensate mode $\Psi_0$. Since this mode depends itself on the number of condensate particles, the product $\sqrt{\langle N_0 \rangle}\Psi_0$ is often called mean field, or order parameter, depending on the community. Note that, within the quantum master equation, the Hartree ansatz is used to quantify the condensate mode assuming all particles to share the same (non-averaged) single particle quantum mode $\Psi_0$. Indeed, we will see that this is an accurate assumption since any term proportional to $g$ in Eq. (2.15) will enter the dynamics and statistics of ultracold quantum gases only negligibly small, if only $a\varrho^{1/3} \ll 1$ (see Part II and Part III).

($N-1$) – here, ($N-1$) is replaced by $N$ for convenience in Eq. (2.15) – other particles with which a condensate particle interacts. From Eq. (2.15), it follows that the condensate mode $\Psi_0(\vec{r},t)$ is a function of the product $gN$. Hence, the condensate wave function $|\Psi_0\rangle$ has a well-defined limit to the Schrödinger wave function $|\chi_0\rangle$ of a particle in an ideal gas in the formal limiting case of weak interactions, $a \to 0^+$ [45].

## 2.4 Theories of condensate growth

There exist different quantum kinetic theories to describe the time evolution of the average condensate fraction during Bose-Einstein condensation. Summarizing their relevant results in a short fashion, we demonstrate that – to our knowledge – none of the theories could so far address the dynamics of the full condensate number distribution *during* Bose-Einstein condensation which highlights the condensate formation process of the atoms in the gas below $T_c$. This is because either the full quantum problem is in most cases impossible to solve numerically, the total particle number is not conserved, or/and condensate formation is studied in terms of the quantum Boltzmann equation. Moreover, we notice that there exists no quantitative master equation theory for *closed and interacting* dilute Bose gases below $T_c$ imposing particle number conservation onto the state of the system.

### 2.4.1 Condensate growth from quantum Boltzmann equation

Many works focus on Bose-Einstein condensation in terms of the quantum Boltzmann equation (QBE) which describes the kinetics of a quantum gas in terms of time dependent particle number occupations $f_{\vec{k}}(t)$. More explicitly, the QBE is given by

$$\frac{\partial f_{\vec{k}}}{\partial t} = \sum_{\vec{l},\vec{m},\vec{n}} \mathscr{C}(\vec{k},\vec{l},\vec{m},\vec{n}) \left[ f_{\vec{m}} f_{\vec{n}} (f_{\vec{l}} + 1)(f_{\vec{k}} + 1) - f_{\vec{k}} f_{\vec{l}} (f_{\vec{m}} + 1)(f_{\vec{n}} + 1) \right] \frac{\delta_{\eta_{\vec{k}} + \eta_{\vec{l}} - \eta_{\vec{m}} - \eta_{\vec{n}}}}{\Delta \eta} , \quad (2.17)$$

## 2.4. THEORIES OF CONDENSATE GROWTH

where $f_{\vec{k}} = f_{\vec{k}}(t)$ denotes the average particle number occupation of a single particle state with energy $\eta_{\vec{k}}$. The Kronecker delta in Eq. (2.17) ensures energy conservation, the choice of the typical spectral energy spacing $\Delta\eta$ depending on the external trapping geometry. The transition rate $\mathscr{C}(\vec{k}, \vec{l}, \vec{m}, \vec{n})$ appearing in Eq. (2.17) is given by

$$\mathscr{C}(\vec{k}, \vec{l}, \vec{m}, \vec{n}) \sim \frac{2\pi g^2}{\hbar} \left| \int d\vec{r}\, \chi_{\vec{k}}^*(\vec{r}) \chi_{\vec{l}}^*(\vec{r}) \chi_{\vec{m}}(\vec{r}) \chi_{\vec{n}}(\vec{r}) \right|^2 , \qquad (2.18)$$

with $\chi_{\vec{k}}(\vec{r})$, the single particle eigenfunctions of a non-interacting gas, and $g$ the two body interaction strength.

The QBE is controversially discussed [57] with regard to its character as a good approximation to the true Hamiltonian dynamics, or, as a defining theory on its own right. From the Hamiltonian point of view, one way to decompose the total Hamiltonian for an interacting gas is $\hat{\mathcal{H}} = \hat{\mathcal{H}}_{\text{ideal}} + \hat{\mathcal{V}}$, with

$$\hat{\mathcal{H}}_{\text{ideal}} = \sum_{\vec{k}} \eta_{\vec{k}} \hat{a}_{\vec{k}}^\dagger \hat{a}_{\vec{k}} \quad \text{and} \quad \hat{\mathcal{V}} = \frac{g}{2} \sum_{\vec{k},\vec{l},\vec{m},\vec{n}} \zeta_{\vec{k}\vec{l}}^{\vec{m}\vec{n}} \hat{a}_{\vec{k}}^\dagger \hat{a}_{\vec{l}}^\dagger \hat{a}_{\vec{m}} \hat{a}_{\vec{n}} , \qquad (2.19)$$

where $\zeta_{\vec{k}\vec{l}}^{\vec{m}\vec{n}} = \int d\vec{r}\, \chi_{\vec{k}}^*(\vec{r}) \chi_{\vec{l}}^*(\vec{r}) \chi_{\vec{m}}(\vec{r}) \chi_{\vec{n}}(\vec{r})$. The interaction term $\hat{\mathcal{V}}$ commutes with $\hat{\mathcal{H}}_0$ only if the energy is balanced, i.e. $\eta_{\vec{k}} + \eta_{\vec{l}} = \eta_{\vec{m}} + \eta_{\vec{n}}$. In this case, the time evolution of the gas is governed by the unitary time propagator $\hat{\mathcal{U}}(t) = \exp[-i\hat{\mathcal{H}}_{\text{ideal}}t/\hbar]\exp[-i\hat{\mathcal{V}}t/\hbar]$. Calculating all different non-vanishing transition rates proportional to $|\langle\{N_{\vec{k}}\}|\hat{\mathcal{U}}(t)|\{N_{\vec{l}}\}\rangle|^2$ between the eigenstates $|\{N_{\vec{k}}\}\rangle$ of $\hat{\mathcal{H}}_{\text{ideal}}$ with respect to the interaction term $\hat{\mathcal{V}}$ to second order (in $g$) with the Fermi golden rule [58], and imposing energy conservation by adding a delta function $\delta(\eta_{\vec{k}} + \eta_{\vec{l}} - \eta_{\vec{m}} - \eta_{\vec{n}})$ to the coupled evolution equations for $f_{\vec{k}}(t) \equiv \langle \hat{a}_{\vec{k}}^\dagger \hat{a}_{\vec{k}} \rangle(t)$ subsequently leads to Eq. (2.17). The problems with the above derivation are:

(i) The eigenvectors of $\hat{\mathcal{H}}_{\text{ideal}}$ are not the same as the eigenvectors of $\hat{\mathcal{H}}$. So, interpreting the expectation value of $\hat{\mathcal{H}}$ as the true energy, the time evolution governed by $\hat{\mathcal{U}}(t)$ cannot lead to any transition between the eigenstates of the Hamiltonian

$\hat{\mathcal{H}}$ for the interacting gas. Thus, the time evolution governed by the QBE (2.17) may therefore not represent the real dynamics of the interacting system, since it assumes transitions and energy conservation with respect to the eigenstates of a non-interacting gas as described by the Hamiltonian $\hat{\mathcal{H}}_{\text{ideal}}$.

(ii) The use of Fermi's golden rule causes a problem, considering that $\hat{\mathcal{H}}_0$ commutes with $\hat{\mathcal{V}}$. Expanding the time propagator $\hat{\mathcal{U}}(t) = \exp[-i\hat{\mathcal{H}}_{\text{ideal}}t/\hbar]\left[\hat{\mathbb{1}} - i\hat{\mathcal{V}}t/\hbar\right]$, and taking the short time limit $t \to 0^+$, the off-diagonal (transition) matrix elements $\langle\{N_{\vec{k}}\}|\hat{\mathcal{U}}(t)|\{N_{\vec{k}'}\}\rangle$ tend to zero, meaning that the instantaneous rates of the collision processes vanish (quantum Zeno effect [57]).

For these reasons, the QBE is often regarded as a phenomenological ansatz rather than being justified from the microscopic point of view. In Chapters 5-7, we will show that – under the inclusion of all two body interactions and accounting for the finite phase coherence time $\tau_{\text{col}}$ between two colliding particles (avoiding the Zeno paradox) – the derived condensate growth equation (6.5) resembles the QBE in Eq. (2.17) for $\vec{k} = 0$ referring to the ground state mode, and assuming a finite with of the $\delta$-function. Our quantum approach does not impose the assumption of microscopic energy conservation, following naturally in the derivation of the master equation. Support to the validity of the QBE is given by the master equation as concerns the use of single particle wave function for non-interacting atoms: we will show that corrections of the atomic single wave functions with respect to atomic interactions (neglected in the microscopic derivation of the QBE) occur indeed as $\mathcal{O}(g^3)$-terms in the time evolution of single particle occupations, scaling as $a\varrho^{1/3} \ll 1$ relatively to the leading order contribution (governed by the QBE). Under the restriction to two body collisions and therefore to contributions proportional to $g^2$ in the limit $a\varrho^{1/3} \ll 1$, the QBE (2.17) is thus perfectly justified from the microscopic point of view, despite lacking the finite width $\Gamma$ of the $\delta$-function, arising from the (finite) spatial phase coherence time between the quantum particles – which has to be included (see Chapter 7).

Since the product of the different occupation numbers on the right hand side of Eq. (2.17) describes the equilibration of the atoms in the gas due to atomic two body

## 2.4. THEORIES OF CONDENSATE GROWTH

collisions, Bose-Einstein condensation is studied within the QKB by regarding the population of the ground state mode $\vec{k} = (0,0,0)$ in Eq. (2.17). A practical problem for simulating the QBE, however, is that the degeneracy of states increases rapidly with increasing energy. For example, the choice $k_B T = 10\hbar\omega$ in an isotropic harmonic trap already limits significantly the numerical speed [59]. Nevertheless, the QBE can be simulated numerically for small total particle numbers ($N \sim 100-1000$) assuming that each occupation number $f_{\vec{k}}$ with arbitrary $\vec{k}$ refering to the same energy depends only on the energy of the state[3] (see Section 2.4.4). The advantage of the condensate growth equation (6.6) is that it can be applied in a large desired range of particle numbers (we probed the range $N = 200$ to $N = 10^6$).

### 2.4.2 Pioneering works of Levich and Yakhot

First investigations on Bose-Einstein condensation have been performed by Levich and Yakhot [60] using the QBE in order to study the dynamics of a gas in a box coupled to a bath of fermionic particles below the critical temperature. In this first study, the authors made the important assertion that there exist two macroscopically distinct stages of condensate formation, the first being a fast equilibration of the gas' high energetic part within a few two body collision times (thermalization), $\tau_{col} \sim 50-100$ ms, and the second stage, the actual formation of the condensate[4] – highlighting a clear separation of the time scale $\tau_{col}$ for thermal equilibration of the non-condensate part of the gas from the time scale $\tau_0$ for Bose-Einstein condensation.

### 2.4.3 Predictions of Kagan, Svistunov and Shlyapnikov

First considerations on Bose-Einstein condensation by Svistunov [61] were also conducted for the simplified case of a Bose gas in a box, replacing the terms

---
[3]so that each state with the same energy has the same occupation number/probability, what represents the ergodic dynamics assumption

[4]In accordance with the separation of time scales found by Holland, Walser and Cooper [59] (see Section 2.4.4).

($f_{\vec{\text{l}}}+1$) by 1, and thus assuming $f_{\vec{\text{l}}} \ll 1$ in Eq. (2.17). Within this approximation, Svistunov was led to an analytical solution for the distribution function $f_0(E,t)$ being a function of energy $E$. According to Svistunov, the distribution $f_0(E,t)$ propagates from high energies to the ground state energy within a time scale $\tau_{\text{col}}$ after it returns to the initial energy region. The predictions of Svistunov correspond to our observation that particles are transported from the non-condensate modes towards the condensate mode (with net positive current towards the condensate mode below $T_c$), until a slow (linear) convergence[5] into a detailed balance particle flow (compare Chapters 6, 8).

Subsequently [61, 62], Kagan, Svistunov and Shlyapnikov studied the dynamics of condensate growth in more detail again showing that there exist two distinct stages of condensate formation: Initially, the non-equilibrium state of the gas rapidly equilibrates and implies the transport of high-energetic particles to the low-energy region, occuring on the average time scale $\tau_{\text{col}}$ of two body collisions in the gas, which is equivalent to the observations of Levich and Yakhot [61, 62] (see Section 2.4.2) and the results of Holland, Walser and Cooper [59, 63] (see Section 2.4.4). Our theory will show that this separation of time scales – theoretically found by Levich et al., Walser et al., and experimentally confirmed by Miesner et al. [64] – enables and justifies the derivation of a Markov quantum master equation. The second step comprises the experimentally observable condensate formation process, where average macroscopic occupation of the ground state mode occurs within a time scale $\tau_0$, which is much longer than the time scale of two body collisions, $\tau_{\text{col}}$.

Kagan, Svistunov and Shlyapnikov cannot give a number for the time scale $\tau_0$, however providing a qualitative understanding of Bose-Einstein condensation. In contrast to $\tau_0$, the time scale $\tau_{\text{col}} \sim 50-100$ ms can be theoretically estimated for a thermal gas (see Chapter 3). A direct monitoring of the full quantum distribution during the second stage of Bose-Einstein condensation in real-time is displayed

---

[5]The slope of the linear growth is time dependent and tends to zero while reaching the equilibrium state.

## 2.4. THEORIES OF CONDENSATE GROWTH

in Chapter 6.

### 2.4.4 Kinetic evolution obtained from Holland, Williams and Cooper

In an early work of Holland, Williams and Cooper [59], the kinetics of condensation formation are studied in a harmonic trap using a simulation procedure of the QBE within the ergodic assumption. Therein, the authors find the characteristic dynamical behavior of exponential condensate growth,

$$f_0(t) = f_0(\infty)\left[1 - e^{-t/\tau_0}\right] , \qquad (2.20)$$

where $f_0(\infty)$ is the equilibrium condensate occupation (which depends on specific parameters such as temperature, trap frequencies, etc.), and $\tau_0 \sim 1\ldots 4$ s, the characteristic time scale of condensate formation, being extracted from the condensate growth curves obtained from the exact numerical propagation of the QBE. The times scales are in agreement with our results in Chapters 6, and Eq. (2.20) qualitatively resembles the condensate growth Eq. (6.6) for the average condensate population as predicted by our master equation theory of Bose-Einstein condensation.

The exact numerical propagation of Eq. (2.17) is possible to be carried out for small particle numbers of the order of $N = 10^2 - 10^3$, entailing the dynamics of the expectation value of the average condensate occupation, $f_0(t)$. As evident from Eq. (2.17), the full quantum state of the Bose gas cannot be reproduced from the QBE. The equilibrium occupations of non-condensate single particle modes $f_{\vec{k}}(\infty)$ are found [59] to be in accordance with Bose-Einstein statistics (including the discrete nature of single particle levels (see Chapter 1)), and hence with the results of our quantum master equation theory (see Chapter 8). Again, the important implication of the QBE for our quantum master equation of Bose-Einstein condensation is the separation of time scales between the thermal equilibration in the gas from the condensate formation time. The time scale for equilibration in the non-condensate, i.e. the high-energetic part of the gas, is [59, 63, 65] of the order of the average time scale for two body collisions, $\tau_{col} \sim 50-100$ ms, whereas condensation formation is predicted to last a few seconds.

### 2.4.5 Stoof's contribution

Stoof [66, 67] adopts the distinct stages of condensation formation in a Keldysh formalism for the condensate mean field. The central result of this theory is that the initial condensate population (nucleation) is due to particle transport from energetically low lying states towards the condensate mode, within a time scale $\tau_{col}$, the relevant time scale for the first equilibration stage of condensation – as in the theory of Kagan, Svistunov and Shlyapnikov (see Section 2.4.3). In addition [68], Stoof presented a Fokker-Planck equation for the distribution function of the condensate mean field, which could highlight the coherent nature of bosonic particles. A numerical solution of this complicated equation, however, has not yet been published (to the best of our present knowledge). Stoof's main assertion concerning the (non-equilibrium) dynamics of the gas is thus that population of the condensate mode occurs at the expense of the energetically low-lying states, a fact that agrees with the condensate formation rates of the present thesis (see Chapters 6, 7). The main contribution to condensate feeding therein is due to the overlap of weakly excited single particle states with the condensate wave function after the initiation of the condensate formation process.

### 2.4.6 Quantum kinetic theory

Quantum kinetic theory (QKT) [17, 18, 69, 70, 71, 72, 73, 74, 75, 76] is the closest non-equilibrium approach to our quantum master equation theory. The results of quantum kinetic theory are summarized for three-dimensional, trapped Bose gases [17, 18], being close to experimental setups and to the case typically considered numerically in Chapters 6 and 8 of this thesis. We point out important conceptual improvements of our theory.

In QKT, single particle states of the Bose gas are devided into a condensate

## 2.4. THEORIES OF CONDENSATE GROWTH

band, $R_0$, and a non-condensate band, $R_\perp$ (we adopt our nomenclature 0 and $\perp$ for condensate and non-condensate, respectively), the first containing all single particle states of which the spectrum is significantly shifted (with respect to the unperturbed one) by the presence of a large average condensate fraction, and the latter consist of all states, where this shift can be neglected. Since the temperature in the experiment is sufficiently large, implying a large non-condensate particle number, the state of the non-condensate is approximated by an *undepleted* thermal mixture,

$$\hat{\rho}_\perp(T) = \mathscr{Z}^{-1} \exp\left[-\beta\left(\hat{\mathcal{H}}_\perp - \mu_\perp N_\perp\right)\right] , \qquad (2.21)$$

where $\hat{\mathcal{H}}_\perp$ is the Hamiltonian of the non-condensate, acting on number states associated with $R_\perp$ only, $N_\perp$ is the (constant) equilibrium particle population in $R_\perp$, $\mu_\perp$ is the time independent (assuming no depletion of the non-condensate) chemical potential of the non-condensate, and $\beta = (k_B T)^{-1}$ is the inverse temperature (of the bath and the locally thermalized non-condensate part of the gas). Hence, the Bose gas in QKT represents an open system, meaning that particles are exchanged with an external reservoir (because $N_\perp$ is fixed and $N_0$ grows), and detailed balance particle flow at equilibrium is particularly reached with the (virtual) external, particle reservoir. In contrast, we assume the particle number to be conserved and respect the depletion of the non-condensate during condensate formation (serving as a finite size thermal environment).

Within QKT, the total state $\hat{\sigma}(t)$ of the Bose gas is supposed to be a product state of the reduced condensate density matrix, $\hat{\rho}_0(t) = \text{Tr}_\perp \hat{\sigma}(t)$, and the time independent thermal state $\hat{\rho}_\perp(T)$ describing the non-condensate, i.e.

$$\hat{\sigma}(t) = \hat{\rho}_0(t) \otimes \hat{\rho}_\perp(T) , \qquad (2.22)$$

even in the presence of particle exchange between condensate and non-condensate.

As we will see in Chapter 5, this is in general not true for a closed gas containing a finite particle number. The derivation of the master equation for various different two body interaction terms follows the standard quantum optical procedure [58, 77]. Two body interactions are distinguished by interactions terms which lead to transport of particles between the bands $R_0$ and $R_\perp$, and those, which leave the occupations in $R_0$ and $R_\perp$ unchanged.

For sufficiently low temperatures and large condensate occupation [17, 18], the condensate band reduces to only one single particle mode, $\Psi_0(\vec{r})$. The bosonic field operator $\hat{\Psi}(\vec{r})$ hence splits into a condensate part, $\hat{\Psi}_0(\vec{r})$, which is (to accuracy $1/N$) determined by the time-independent solution of the Gross-Pitaesvkii equation with $N$ particles occupying the single particle ground state mode $\Psi_0(\vec{r})$ (see Eq. (2.15) of Chapter 3) and by the non-condensate field $\hat{\Psi}_\perp(\vec{r})$, expanded in the momentum basis, which spans the single particle subspace orthogonal to the condensate mode.

As mentioned before, the non-condensate is treated as an undepleted thermal mixture of non-condensate particles, see Eqs. (2.21, 2.22), with a linearized non-condensate Hamiltonian using a Bogoliubov transformation [17, 18]. An equation for condensate growth is obtained by evaluating the different terms for two body interaction processes (see also Chapter 4), and by neglecting terms which act within $R_0$ and $R_\perp$ only, as well as terms which account for pair processes, i.e., processes which create and annihilate two condensate particles simultaneously. Those processes are found to change the condensate dynamics only slightly, confirmed in Chapter 7 of this thesis, showing that they occur as off-resonant (not energy conserving) in the dynamical evolution of the gas.

All together, this leads to the following equation for the diagonal elements $p(N_0, t) = \langle N_0 | \hat{\rho}_0(t) | N_0 \rangle$ of the reduced condensate density matrix [17, 18]:

$$\frac{\partial p(N_0, t)}{\partial t} = 2N_0 \lambda^+(N_0 - 1, T) - 2(N_0 + 1)\lambda^+(N_0, T)p(N_0, t) \\ + 2(N_0 + 1)\lambda^-(N_0 + 1, T)p(N_0 + 1) - 2N_0 \lambda^-(N_0, T)p(N_0, t) \ . \quad (2.23)$$

## 2.4. THEORIES OF CONDENSATE GROWTH

The essential physics of Bose-Einstein condensation lies in the condensate feeding and loss rates, $\lambda^+(N_0, T)$ and $\lambda^-(N_0, T)$. Within quantum kinetic theory, they are given by

$$\lambda^\pm(N_0, T) = \frac{\hbar^2 a^2}{2m^2 \pi^3} \int d\vec{r} \int\int\int\int d\vec{k}\, d\vec{l}\, d\vec{m}\, d\vec{n}\, \mathcal{F}^\pm(\vec{r}, \vec{k}, \vec{l}, \vec{m}, \vec{n}) \delta(\Delta\omega(\vec{r}) - \omega) \delta(\vec{k} + \vec{l} - \vec{m} - \vec{n}) \ , \quad (2.24)$$

where

$$\mathcal{F}^+(\vec{r}, \vec{k}, \vec{l}, \vec{m}, \vec{n}) = f_{\vec{l}}(\vec{r}) f_{\vec{m}}(\vec{r}) [1 + f_{\vec{n}}(\vec{r})] \frac{1}{(2\pi)^3} \int d\vec{v}\, \Psi_0(\vec{r} + \frac{\vec{v}}{2}) \Psi_0(\vec{r} - \frac{\vec{v}}{2}) e^{i\vec{k}\cdot\vec{v}} \ , \quad (2.25)$$

for feeding processes, and

$$\mathcal{F}^-(\vec{r}, \vec{k}, \vec{l}, \vec{m}, \vec{n}) = [f_{\vec{l}}(\vec{r}) + 1][f_{\vec{m}}(\vec{r}) + 1] f_{\vec{n}}(\vec{r}) \frac{1}{(2\pi)^3} \int d\vec{v}\, \Psi_0^*(\vec{r} + \frac{\vec{v}}{2}) \Psi_0(\vec{r} - \frac{\vec{v}}{2}) e^{-i\vec{k}\cdot\vec{v}} \ , \quad (2.26)$$

for losses of condensate particles. Here, $f_{\vec{k}}(\vec{r})$ are occupation numbers of non-condensate particles with momentum $\vec{k}$ at position $\vec{r}$, $\Psi_0(\vec{r})$ is the condensate wave function, $\omega = \omega_{\vec{k}}(\vec{r}) = \hbar \vec{k}^2 / 2m + V_{ext}(\vec{r})/\hbar$, given the external confinement of the three-dimensional harmonic trapping potential, $V_{ext}(\vec{r}) = m/2(\omega_x x^2 + \omega_y y^2 + \omega_z z^2)$. $\Delta\omega(\vec{r}) = \omega_{\vec{l}}(\vec{r}) + \omega_{\vec{m}}(\vec{r}) - \omega_{\vec{n}}(\vec{r})$ is the energy difference of a particular two body collision process, respectively, and the $\delta$ symbol is a Kronecker delta, which doesn't account for the finite spatial coherence time of the interacting particles. As the non-condensate is described as an undepleted thermal mixture at temperature $T$, the average non-condensate particle number occupations $f_{\vec{k}}(\vec{r})$ are given by

$$f_{\vec{k}}(\vec{r}, \vec{k}) = \frac{1}{e^{\beta[\hbar\omega(\vec{r},\vec{k}) - \mu_\perp]} - 1} \ , \quad (2.27)$$

depending on the coordinate $\vec{r}$ and the momentum $\vec{k}$. It is evident from the explicit expressions of the transition rates in Eq. (2.24) that evolution Eq. (2.23) is highly complex to solve numerically, because of the extensive momentum and spatial integrals, leading to an exponential growth of the internal degrees of freedom. Hence, the master Eq. (2.23) was not solved in a numerically exact way [17, 18]. Within our theory, we find the formally equivalent equation (6.3), the master equation of Bose-Einstein condensation, which is numerically accessible — due to the choice of our single particle basis and a perturbation theory for single particle wave functions in the small parameter $a\varrho^{1/3} \ll 1$ (see Chapters 7 and 6).

To obtain condensate formation rates approximately, it is firstly assumed in QKT that the ground state condensate wave function $\Psi_0(\vec{r})$ is sharply peaked at the center of the trap (an assumption which becomes accurate, if the interactions are weak and the temperature is low), so that the spatial dependence of the occupation numbers $f(\vec{r},\vec{k})$ are ignored and replaced by the value $f(0,\vec{k})$ at the center of the trap. The latter are approximated/replaced by the Maxwell-Boltzmann form $f(0,\vec{k}) \to f(\vec{k}) \approx e^{-\beta(\hbar\omega(\vec{k})-\mu_\perp)}$, being assumed to be negligible as compared to unity, $f(\vec{k})+1 \simeq 1$ (an approximation which looses its validity in the highly quantum degenerate limit of low temperatures). The condensate feeding rate is finally evaluated [17, 18] to be

$$\lambda^+(N_0,T) \approx \frac{4ma^2}{\pi\hbar^3\beta^2}e^{2\beta\mu_\perp}[\beta\mu_0(N_0)K_1(\beta\mu_0(N_0))] \,, \tag{2.28}$$

where $K_1$ is a Bessel function [40]. The loss rate is acquired from a balance condition [17, 18],

$$\lambda^+(N_0,T) = e^{\beta(\mu_\perp-\mu_0(N_0))}\lambda^-(N_0,T) \,. \tag{2.29}$$

To obtain the feeding rate in Eq. (2.28), the condensate wave function and the corresponding chemical potential is obtained by neglecting the Laplacian term of the Gross-Pitaevskii equation [16], see Eq. (2.15), with respect to the interaction energy, called "Thomas-Fermi approximation", valid for strongly interacting gases.

## 2.4. THEORIES OF CONDENSATE GROWTH

From our point of view, this varies with the second order Born approximation needed to derive the master equation [17, 18, 77] and to the previous assumptions for approximating $\Psi_0(\vec{r}) \simeq \Psi_0(0)$.

In contrast, the quantum master equation of this thesis employs the diluteness of Bose-Einstein condensates to derive the time scales for condensate formation neglecting terms of the order $a\varrho^{1/3} \ll 1$ (see Chapter 6) in accordance with the second order iteration of the quantum master equation. Notably, the condensate feeding and loss rates of QKT are independent of the total number of particles in the system, and the non-condensate chemical potential is constant. One of the purposes of this thesis is to show that the essential dynamics of Bose-Einstein condensation is due to the change of this chemical potential accounting for the dynamics of the non-condensate environment. The arising non-condensate number fluctuations in number representation are deeply related to the spatial quantum coherence of the gas particles. The circumvention of the above approximations for the transition rates is hence conceptually important, and finally leads to a direct monitoring of the full condensate and non-condensate quantum distribution functions during condensate formation, spelling out the interplay of spatial quantum coherences and quantum number fluctuations (see Chapters 8 and 7).

The eigenvalue of the Gross-Pitaevskii equation $\mu_0(N_0)$ in QKT is evaluated within the Thomas-Fermi approximation, given [16] by

$$\mu_0(N_0) = \left(15a\omega_x\omega_y\omega_z m^{1/2}\hbar^2/2^{5/2}N_0\right)^{2/5} . \tag{2.30}$$

As the Bessel function $K_1$ is close to unity [17, 18] in Eq. (2.28), the rates are approximately independent (or only slightly dependent) on the condensate particle number (which is assumed to be of the order of the total particle number $N$).

To study condensate formation within QKT in a quantitative manner, a "simple growth equation" was derived from the master Eq. (2.23) under the neglect of particle number fluctuations, i.e. the finite width of the condensate number distribution

$p(N_0, t)$ [17, 18, 19], leading to

$$\frac{\partial N_0(t)}{\partial t} = 2\lambda^+(N_0, T)\left[\left(1 - e^{\beta(\mu_0(N_0) - \mu_\perp)}\right) N_0(t) + 1\right],\qquad(2.31)$$

with the condensate feeding rate $\lambda^+(N_0, T)$ in Eq. (2.28). This equation was simulated to study condensate growth [17, 73], typically leading to a S-shaped condensate growth curve, as shown in Fig. 6.3, complying with approximately the same final saturation behavior as the exponential law in Eq. (2.20). The relation of the growth Eq. (6.6) arising from the master equation to the growth Eq. (2.31) is discussed in detail in Chapter 6.

Since the gas is not closed and the state is assumed to factorize, QKT can evidently not yield an equilibrium steady state of a closed gas of exactly $N$ number of particles.

## Survey: Which current aspects can we adopt to monitor the many body dynamics during Bose-Einstein condensation?

In essence, we can learn from, and keep the following aspects from the short introduction in the previous two chapters:

**(i)** *Fundamental aspects*: Bose-Einstein condensation occurs, if the thermal de Broglie wave length of the bosonic atoms is larger than their mean average distance. In this quantum degenerate regime, particles have to be treated as indistinguishable (see Section 1.3), in order to observe Bose-Einstein condensation.

**(ii)** *Statistical aspects*: The grand canonical and the canonical ensemble are no longer equivalent below the critical temperature, even in the thermodynamic limit (see Section 1.5). Comparison of predictions for condensate number expectation values and variances indicate that the most reasonable statistical predictions are governed by the equilibrium (thermal) state of the canonical ensemble – i.e.,

## 2.4. THEORIES OF CONDENSATE GROWTH

assuming a fixed total particle number in the gas. We will hence consider a conserved total number of particles in the Bose gas for the derivation of the quantum master equation.

(iii) **Condensate mode**: We use the Gross-Pitaesvkii equation to define a condensate mode (see Section 2.3). We will see in Part II of this thesis that the terms in the Gross-Pitaevskii equation proportional to $g$ enter the many body dynamics of dilute atomic gases only with negligible terms of the order $a\varrho^{1/3} \ll 1$. An important implication from state-of-the art experiments on Bose-Einstein condensation (see Section 1.4) is the existence of this naturally small parameter, $a\varrho^{1/3}$, in most of the available Bose-Einstein condensates. This parameter can be used to expand the condensate wave function perturbatively, leading to a numerically efficient and quantitatively accurate monitoring of Bose-Einstein condensation (in Part III).

(iv) **Separation of time scales**: Concerning theories for average condensate growth (see Section 2.4), we use the clear separation of the time scale for rapid non-condensate thermalization from the time scale for condensate formation. This is the fundamental background for a quantum master equation ansatz for the condensate part which undergoes its time evolution in the presence of the non-condensate environment. As already mentioned, the separation of time scales has also been confirmed experimentally (see Section 2.4.3).

(v) **Condensate formation:** An evolution equation for condensate growth is meant to predict a typical S-shape behavior and condensate formation times of the order of a few seconds (see Section 2.4). In order to monitor the full quantum distributions of the Bose gas and to deduce reliable predictions for condensate formation times, we have to well include the wave nature of the particles, i.e. the spatial variations of the single particle wave functions, as well as the finite spatial phase coherence time between the interacting particles (leading to a finite energy uncertainty).

# Part II

# QUANTUM MASTER EQUATION OF BOSE-EINSTEIN CONDENSATION

*Reality is what we can calculate. An experimentalist would probably replace the word "calculate", by "measure".*

David Gross, Nobel Lecture December 8, 2004

Chapter 3

# Concepts, basic assumptions and validity range

This chapter briefly summarizes the novel conceptual parts of the quantum master equation theory of Bose-Einstein condensation developed throughout Chapters 4-8. The separation of time scales between condensate formation and non-condensate thermalization in dilute atomic gases is discussed in Section 3.1, enabling the derivation of a Markov quantum master equation for the reduced condensate state in the presence of the non-condensate environment. In Section 3.2, we collect considerations on two body interactions, particle number conservation, rapid non-condensate thermalization and the depletion of the non-condensate thermal environment during condensate formation, required in order to conceptually improve the existing theories of condensate growth summarized in Section 2.4. The $N$-body Born-Markov ansatz for a dilute Bose-Einstein condensate motivated by the separation of time scales is explained in Section 3.3. The validity range of the master equation theory of Bose-Einstein condensation is justified for the case of sufficiently dilute atomic gases, $a\varrho^{1/3} \ll 1$, as argued in Section 3.4.

## 3.1 Motivation for master equation: Separation of time scales

The fundamental property behind the derivation of the master equation is the separation of time scales between non-condensate thermalization and condensate formation. A sketch of the physical situation to be modeled is displayed in Fig. 3.1: As the Bose gas is cooled below the critical point with an evaporative cooling cycle, a condensate appears at the center of the trap, arising from the residual gas of non-condensate particles [78].

Considering one such evaporative cooling step (one sequence in Fig. 3.1), the typical time scale of condensate formation is given by $\tau_0 \sim 1000-4000$ ms [17, 18, 59, 64, 76]. On the other hand, one observes that equilibration within the gas of non-condensate particles occurs on a much time scale, $\tau_{col} \approx 10-100$ ms [47, 51, 59, 64] (see also Appendix A.5). Thus, there exists a clear separation of the time scale for condensate formation from the equilibration time within the non-condensate,

$$\tau_{col} \ll \tau_0 \ . \tag{3.1}$$

Conceptually, this physical separation of time scales allows us to trace out the non-condensate, and thereby to derive a master equation for the reduced condensate density matrix. More specifically, the separation of time scales is expressed in two different formal assumptions required for the derivation of the master equation: (i) we describe the non-condensate as a diagonal thermal state for each given number occupation of the condensate and non-condensate subsystem employing particle number conservation ($N$-body Born ansatz, see Section 3.3.2), and (ii) we suppose that spatial phase coherences between the condensate and the non-condensate particles decay rapidly within the time scale $\tau_{col}$ (Markov assumption in a Bose-Einstein condensate, see Section 3.3.3) faster than the finite time resolution $\Delta t$ (coarse-grained rate of variation [20]) – yielding the Born-Markov ansatz for a Bose-Einstein condensate of $N$ particles. This "$N$-body Born-Markov ansatz" is discussed in more detail in Section 3.3 leading to a closed, time local quantum master equation of Lindblad type for the reduced condensate density matrix.

## 3.2. MODELING OF MANY PARTICLE DYNAMICS

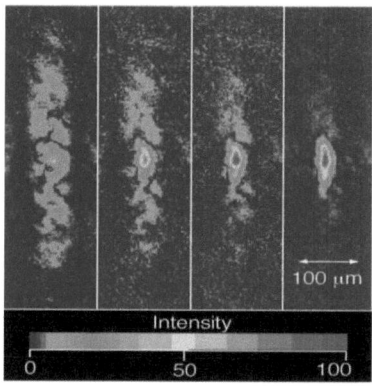

Figure 3.1: *Evaporative cooling cycles to achieve Bose-Einstein condensation [78]. The non-condensate, corresponding to the low-intensity fraction (green area – distributed around the center of the atomic cloud), surrounds and interacts with the condensate, which is shown as the high-intensity part (yellow-red area – center of the atomic cloud) of the backscattered light. Our goal is to model one evaporative cooling step (three different steps are displayed in sequences from left to right in the figure), where the gas forms a Bose-Einstein condensate due to atomic collisions [60, 64] within a typical formation time of $\tau_0 \sim 1000 - 4000$ ms [64, 76].*

Since the time scale $\tau_0$ for the overall process of interest – Bose-Einstein condensation – is much larger than $\Delta t$, a coarse-grained description for the time evolution is appropriate to describe also the instantaneous rate of condensate formation with a quantum master equation. In the following, the most important concepts to describe the many particle dynamics in the gas are summarized, before the $N$-body Born-Markov ansatz for a Bose-Einstein condensate is explained and justified.

## 3.2 Modeling of many particle dynamics

In this section, we summarize the strategy for modeling the $N$-body dynamics during Bose-Einstein condensation. For technical details, consult chapters 4-8.

## 3.2.1 Two body interactions in dilute gases

In sufficiently dilute atomic gases, it is justified to account only for two body interaction processes [79] in the Bose gas. This amounts to replace the exact interaction potential $V_{int}(\vec{r}_1,\ldots,\vec{r}_N)$ by the contact potential $V(\vec{r}_i - \vec{r}_j) = g\delta(\vec{r}_i - \vec{r}_j)$ introduced in Section 2.1,

$$V_{int}(\vec{r}_1,\ldots,\vec{r}_N) \to \frac{1}{2}\sum_{i\neq j} V(\vec{r}_i - \vec{r}_j) + \mathcal{O}(V^{(3)}) \;. \tag{3.2}$$

All two body interactions are described by *one* effective interaction strength, $g = 4\pi\hbar^2 a/m$, given [16] in terms of the s-wave scattering length $a$:

$$V(\vec{r}_i - \vec{r}_j) = \frac{4\pi\hbar^2 a}{m}\delta(\vec{r}_i - \vec{r}_j) \;. \tag{3.3}$$

Using the formalism of second quantization, the N-body Hamiltonian is given by

$$\hat{\mathcal{H}} = \int_{\mathscr{C}} d\vec{r}\, \hat{\Psi}^\dagger(\vec{r}) \left[ -\frac{\hbar^2 \vec{\nabla}^2}{2m} + V_{ext}(\vec{r}) \right] \hat{\Psi}(\vec{r}) + \frac{g}{2}\int_{\mathscr{C}} d\vec{r}\, \hat{\Psi}^\dagger(\vec{r})\hat{\Psi}^\dagger(\vec{r})\hat{\Psi}(\vec{r})\hat{\Psi}(\vec{r}) \;, \tag{3.4}$$

where $\hat{\Psi}(\vec{r})$ denotes the second quantized bosonic field, and $\mathscr{C} \subset \mathbb{R}^3$ the volume of the trap (see Section 2.2). Higher order corrections to the s-wave scattering approximation are negligibly small, if the gas is dilute, $a\varrho^{1/3} \ll 1$ (see Section 2.1).

## 3.2.2 Condensate and non-condensate subsystems

Motivated by the separation of time scales (see Section 3.1), we split the N-particle Bose gas into a condensate and a non-condensate subsystem. For this purpose, the second quantized field is decomposed into

$$\hat{\Psi}(\vec{r}) = \Psi_0(\vec{r})\hat{a}_0 + \hat{\Psi}_\perp(\vec{r}) \;. \tag{3.5}$$

## 3.2. MODELING OF MANY PARTICLE DYNAMICS

Here, $\Psi_0(\vec{r})$ denotes the condensate wave function, which we quantify by the Gross-Pitaevskii equation (see Eq. (2.16) of Chapter 2.3). The operator $\hat{a}_0$ annihilates a particle in the condensate mode. On the other hand, $\hat{\Psi}_\perp(\vec{r}) = \sum_{k \neq 0} \Psi_k(\vec{r}) \hat{a}_k$ denotes the non-condensate field operator, with annihilation operators $\hat{a}_k$ of the single particle modes $\Psi_k(\vec{r})$, which are by definition orthogonal to the condensate mode $\Psi_0(\vec{r})$ (see Section 3.2.3 and Chapter 4).

Corresponding to the splitting of the second quantized field in Eq. (3.5), the Hamiltonian in Eq. (3.4), including two body interactions, falls into

$$\hat{\mathcal{H}} = \hat{\mathcal{H}}_0 + \hat{\mathcal{H}}_\perp + \hat{V}_{0\perp} , \qquad (3.6)$$

where $\hat{\mathcal{H}}_0$ and $\hat{\mathcal{H}}_\perp$ denote the condensate and the non-condensate Hamiltonian[1], respectively, and $\hat{V}_{0\perp}$ the various two body interaction processes between condensate and non-condensate. The latter can be classified as single particle events ($\Delta N_0 = -\Delta N_\perp = \pm 1$, labeled by ⤳), pair events ($\Delta N_0 = -\Delta N_\perp = \pm 2$, labeled by ⤳⤳) and scattering events ($\Delta N_0 = \Delta N_\perp = 0$, labeled by ↻), according to the net exchange of condensate particles $\Delta N_0$ per two body interaction process.

### 3.2.3 Thermalization in the non-condensate

To model the thermalization process arising from self-interactions in the non-condensate during condensate formation, we introduce the following approximation: we replace the non-condensate Hamiltonian $\hat{\mathcal{H}}_\perp$ by its linearized version, and account for thermalization within the non-condensate by coupling the non-condensate to an external heat reservoir of fixed temperature $T$. Thereby, we arrive at the picture given in Fig. 3.2, demonstrating the modeling of the microscopic many particle dynamics and the subdivision of the Bose gas into the subsystems condensate and non-condensate.

---
[1]The Hamiltonian parts $\hat{\mathcal{H}}_0$ and $\hat{\mathcal{H}}_\perp$ are time independent.

# Chapter 3. CONCEPTS, BASIC ASSUMPTIONS AND VALIDITY RANGE

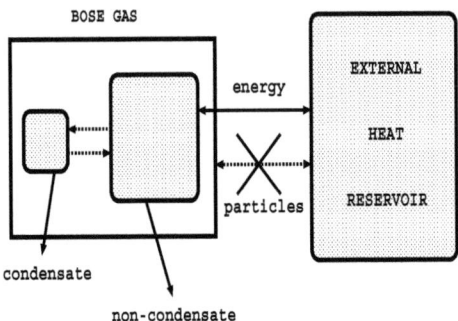

Figure 3.2: *Representation of microsopic many particle dynamics. The total number of atoms in the Bose gas is fixed to N and conserved during condensate formation. Atomic collisions within the non-condensate are modeled by coupling the non-condensate part of the gas to a heat reservoir, which has a fixed temperature T. The condensate part is initially not in detailed particle balance with the non-condensed fraction, and both systems undergo a net exchange of particles, induced by atomic two body collisions between condensate and non-condensate atoms, which are fully taken into account in the derivation of the master equation. Finally, an equilibrium steady state of the gas is reached corresponding to the appearance of the condensate after one evaporative step in Fig. 3.1 below $T_c$, exhibiting detailed balance particle flow between condensate and non-condensate.*

It is important to note that only energy, but no particles are exchanged with the thermal environment, since any two body interaction event in the Bose gas leaves the total particle number invariant. Hence, the particle number $N$ of the Bose gas remains conserved (after the completion of the evaporative shock cooling cycle) throughout the whole condensation process.

Furthermore, we note that only the non-condensate but not the condensate subsystem is directly coupled to the environment (because of the separation of time scales), see Fig. 3.2. Thus, the condensate subsystem is only coupled to the non-condensate via the interaction Hamiltonian $\hat{V}_{0\perp}$ – even in the limit of very weak interactions, $a \to 0^+$, i.e. the steady state depend on $\hat{V}_{0\perp}$. In contrast, the standard thermodynamical approach in the canonical ensemble (compare Section 1.5) assumes the coupling of the whole system (condensate and non-condensate) to the thermal heat bath, see Fig. 1.4. The corresponding thermal equilibrium state is independent of $\hat{\mathcal{V}}_{0\perp}$ in the limit $a \to 0^+$, where it reduces to the thermal equilibrium state of an ideal gas. These considerations indicate that there is, apriori, no guaranty that, according to the model shown in Fig. 3.2, the Bose gas approaches finally a thermal equilibrium state. We therefore compare the steady state of the quantum master equation to the canonical Boltzmann thermal state of an ideal quantum gas in Eq. (1.23) (see Chapter 8), showing that this steady state is a thermal Boltzmann state under the Markov dynamics assumption (which applies, if the interacting particles exhibit a finite spatial phase coherence time, $\hbar \tau_{\text{col}}^{-1} \beta \ll 1$).

## 3.3 N-body Born-Markov ansatz

Here, we explain the Born-Markov ansatz for a Bose-Einstein condensate of exactly $N$ particles under the inclusion of two body interactions, arising from the coupling of the non-condensate to a thermal bath (see chapter 3.2.3) and the separation of time scales discussed in Section 3.1.

### 3.3.1 General Born-Markov ansatz

The basic, general assumption for the derivation of a quantum Markov master equation [19] is the neglect of memory effects of the system's past onto its present time evolution, due to the rapid decay of memory effects in the external environment. In other words, the total state of the system plus environment at any time $t$ is considered to be approximately characterized in terms of the reduced state of the system at the same time $t$, independent on its previous history.

Typically, one therefore assumes initially a product state between system and environment, the latter being in (local) thermal equilibrium, $\hat{\rho}(0) = \hat{\rho}_S(0) \otimes \hat{\rho}_E(T)$. Furthermore, the environment is supposed to remain in the same thermal state also during the subsequent time evolution, i.e., the total state is $\hat{\rho}(t) = \hat{\rho}_S(t) \otimes \hat{\rho}_E(T) + \delta\hat{\rho}(t)$

with negligible deviation $\delta\hat{\rho}(t)$ from a product state of system and environment.

In the present case of the condensate as a system and the non-condensate as an environment part, however, the total $N$–body state $\hat{\sigma}^{(N)}(t)$ of the gas is actually *not* a product state between a condensate and non-condensate density matrix,

$$\hat{\sigma}^{(N)}(t) \neq \hat{\rho}_S(t) \otimes \hat{\rho}_E(T) \, . \tag{3.7}$$

This is due to correlations originating from particle number conservation: obviously, if there are $N_0$ populating the condensate mode, this determines the number of non-condensate particles as $(N-N_0)$. However, it turns out that this is not excluding the derivation of a Markov quantum master equation, as long as the total state $\hat{\sigma}^{(N)}(t)$ is completely determined by the reduced state of the condensate, $\hat{\rho}_0(t) = \text{Tr}_\perp \hat{\sigma}^{(N)}(t)$, such as in Eq. (3.8).

### 3.3.2 Born ansatz for gases of fixed particle number

For this purpose, the constraint of particle number conservation has to be employed when approximating the non-condensate by a thermal state. Indeed, since the particle exchange with the environment is not allowed, thermal equilibration can only occur within subspaces of fixed particle number. Assuming, in addition, that any coherences between subspaces of different particle number are destroyed by the thermalization process leads us to the following ansatz for the total $N$-body state of the gas:

$$\hat{\sigma}^{(N)}(t) \simeq \sum_{N_0=0}^{N} p_N(N_0, t) |N_0\rangle\langle N_0| \otimes \hat{\rho}_\perp(N-N_0, T) \, , \tag{3.8}$$

where $|N_0\rangle\langle N_0|$ denotes a state of $N_0$ particles populating the condensate mode $\Psi_0(\vec{r})$ (see Section 3.2.2), weighted with the positive probability distribution $p_N(N_0, t)$. The states $\hat{\rho}_\perp(N-N_0, T)$ in Eq. (3.8) represent thermal mixtures of $(N-N_0)$ non-condensate particles,

$$\hat{\rho}_\perp(N-N_0,T) = \hat{\mathcal{D}}_{N-N_0} \frac{e^{-\beta\hat{\mathcal{H}}_\perp}}{\mathcal{Z}(N-N_0)} \hat{\mathcal{D}}_{N-N_0}, \qquad (3.9)$$

leading to non-condensate single particle occupations (see Appendix A.3):

$$f_k(N-N_0,T) \equiv \text{Tr}_\perp\{\hat{a}_k^\dagger \hat{a}_k \hat{\rho}_\perp(N-N_0,T)\} = \frac{1}{e^{\beta(\epsilon_k - \mu_\perp(N-N_0,T))} - 1}. \qquad (3.10)$$

The temporal change of single particle occupations while reaching the Bose condensed phase can thus be described in terms of the function $\mu_\perp(N-N_0,T)$, which normalizes non-condensate single particle occupations to $(N-N_0)$ particles, given that $N_0$ particles populate the condensate mode. Due to particle number conservation, $\mu_\perp(N-N_0,T)$ obeys the closed equation:

$$\sum_{k\neq 0} \frac{1}{e^{\beta(\epsilon_k - \mu_\perp(N-N_0,T))} - 1} = (N-N_0). \qquad (3.11)$$

The $N$-body Born ansatz in Eq. (3.8) furthermore allows us to express the entire state of the Bose gas as a function of the reduced condensate state defined by the condensate number distribution $p_N(N_0,t)$, thereby enabling the derivation of a closed evolution equation for the latter under the inclusion of all two body interaction processes between the condensate and the non-condensate particles. This equation, however, may still contain memory effects. In order to get a time local Markov quantum master equation of Lindblad type, the rapid decay of spatial phase coherences (Markov assumption) has to be adopted.

### 3.3.3 Markov approximation for a Bose-Einstein condensate

Any collision event of two particles in a *quantum gas* produces spatial phase coherences (coherent coupling of single particle wave functions in position space) between them. The thermalization process of the non-condensate environment erases these coherences, thus providing the rapid decay of correlations between system and environment required for the Markov dynamics assumption in a Bose-Einstein condensate. The corresponding decay rate can be estimated [47, 51, 59, 64] to be $\tau_{col} = (\varrho a^2 \bar{v})^{-1} \sim$ ms, with $\bar{v} = \sqrt{3k_B T/m} \sim$ cm/s (see also Section 2.4 for a detailed discussion), and implemented in our theory by assuming an irreversible Gaussian decay of the non-condensate correlation functions within the time scale $\tau_{col}$.

Microscopically, this decay process can be understood as due to the fact that a non-condensate particle with which a condensate particle obeys phase coherence has a higher probability to collide (randomly) with many other non-condensate particles than subsequently colliding with a condensate particle again. Hence, spatial phase coherences between condensate and non-condensate particles are rapidly destroyed by non-condensate thermalization.

## 3.4 Limiting cases and validity range

Now, the dilute gas limit, the perturbative limit of very weakly interacting gases and the thermodynamic limit are differentiated and discussed [10].

### 3.4.1 Dilute gas condition

The dilute gas condition is defined by the requirement that the gas parameter $a\varrho^{1/3} \ll 1$, where $\varrho$ is the atomic density and $a$ is the s-wave scattering length. As a matter of fact, this condition theoretically arises from the Born approximation and therefore from the use of an s-wave scattering length $a$ (see Section 2.1) to describe two body interactions. This reduces the validity range of the master equation theory to

$$\xi = a\varrho^{1/3} \ll 1 \,. \tag{3.12}$$

Since the condition $a\varrho^{1/3} \ll 1$ implies $a^3\varrho \ll 1$ (but not vice versa), it justifies as well the neglect of three body and higher order atomic collisions [59]. Hence, $\xi$ is considered to be the small parameter of our theory, which is indeed the case for many experiments treating dilute atomic gases where typically $\xi \sim 10^{-2}$ [15, 23, 24, 32, 64] (see Section 1.4).

### 3.4.2 Perturbative limit

For numerical simplicity, quantitative numerical calculations throughout the thesis are restricted to the weakly interacting case, formally indicated by the limit $\xi \to 0^+$ with $\varrho = $ const and $a \neq 0$. Within this limit, a perturbation theory for single particle wave functions shows that the transition rates between condensate and non-condensate to the lowest non-vanishing order are still proportional to $a^2$, however being fully quantified by the Schrödinger equation (see Chapter 7).

The perturbative limit yields equilibrium distributions for weakly interacting gases given by the ratio of the leading order terms for two body transition rates. However, these do still contain the specific nonlinearity of two body collisions which lead to condensate formation, in contrast to a thermal state of an ideal gas. The steady state distributions of the master equation are therefore compared to the thermodynamic prediction of the canonical ensemble (in Section 1.5.2) for non-interacting gases.

### 3.4.3 Thermodynamic limit

It is important to distinguish the dilute gas limit from the thermodynamic limit [10], which is specified by the condition that

$$N \to \infty \quad \text{with} \quad T_c = \text{const}, \tag{3.13}$$

being in general different from the dilute gas limit in Eq. (3.12). The thermodynamic limit accounts for the asymptotic of large particle numbers and large traps at constant density, independent of the actual value of the interaction strength $g$ and the atomic density $\varrho$. In contrast, the dilute gas and the perturbative limit employ that the effective interaction range of the particles is much smaller than the average distance between them.

### 3.4.4 Semiclassical limit

The semiclassical limit assumes the single particle spectrum to be quasi continuous (see chapter 1), replacing the sums over states by an integral to calculate thermodynamic properties (such as the non-condensate number occupation), smearing out the details (in particular the degeneracy of single particle states) of the single particle spectrum. This limit results in a (positive) shift of the critical temperature for Bose-Einstein condensation with respect to the prediction for $T_c$ including the discreteness of the single particle energies (see Chapters 1 and 8). We will see in Chapter 8 that the shift of the critical temperature does not originate from the neglect of the single particle ground state energy (zero-point motion [15]), but is a result of the quasi continuum approximation.

### 3.4.5 Physical realization of limiting cases

The semiclassical and the thermodynamic limits are difficult to realize exactly [15, 80], being likely to reflect the physics of bosonic gases with particle numbers of the order of Avogadro's number, $N \sim 10^{23}$, a fact that motivates the development of quantitatively accurate theories for mesoscopic quantum gases.

The situation is different for the perturbative and the dilute gas limit valid in particular for small atomic gases, which are used throughout the thesis: For a three-dimensional (isotropic) harmonic trap, the formal perturbative limit, $\xi \to 0^+$ with $\varrho = $ const. and $a \neq 0$, is well realized, if the single particle energy ($\hbar\omega$) exceeds the condensate interaction energy ($gN_0/L^3$), with $L = \sqrt{\hbar/m\omega}$ as the unit length of the harmonic oscillator (see Section 1.5). In this case, $a/LN_0 \ll 1$ is a leading

## 3.4. LIMITING CASES AND VALIDITY RANGE

condition for the applicability of perturbative transition rates, being realized with small atomic samples of the order of a few hundred of atoms (or using Feshbach resonances to reduce the atomic interaction strength [15]). Quantitative predictions of the perturbative limit may as well apply in the dilute gas regime as indicated by the comparisons to experimental condensate formation times in Chapter 6. This statement, however, cannot be proven analytically.

The dilute gas limit – the validity range of the master equation theory of Bose–Einstein condensation – is satisfied in most experiments with currently available alkali species [15] (see Section 1.4). A leading condition for its applicability is therefore $a\varrho^{1/3} \ll 1$.

Chapter 4

# Quantized fields, two body interactions and Hilbert space

The physical separation of time scales between non-condensate thermalization and condensate growth (see Section 3.3.1) motivates a formal decomposition of the gas into a condensate "system" part and a non-condensate "environment" part. In this chapter, we establish the algebraic background for the derivation of the master equation. To this end, the condensate mode is defined in Section 4.1. In Section 4.2, this definition is used to separate the second quantized condensate field from the non-condensate field. Correspondingly, the interaction term of the $N$-body Hamiltonian in Eq. (3.4) yields formally nontrivial two body interactions between condensate and non-condensate besides the Hamiltonian parts describing the coherent time evolution in the master equation later on. Interactions in a Bose-Einstein condensate fall into three different physically motivated classes of two body vertices: single particle, pair and scattering events. Finally, diagonalization of the Bose gas' non-condensate Hamiltonian leads to a single particle basis in Section 4.3, defining the underlying single particle Hilbert space as well as the many particle Fock-Hilbert space for a gas of $N$ indistinguishable particles. Those are analyzed in Section 4.4.

## 4.1 Definition of the condensate

For a gas of $N$ interacting bosonic particles in an arbitrary external trapping potential, the condensate and the non-condensed part can be defined [81] with the help of the single particle density matrix of Eq. (1.28),

$$\hat{\rho}^{(1)}(t) = N\text{Tr}_{2,\ldots,N}\hat{\sigma}^{(N)}(t) \,, \tag{4.1}$$

where $\hat{\sigma}^{(N)}(t)$ is the exact N-body state of the interacting system at time $t$, and $\hat{\rho}^{(1)}(t)$ is normalized to $N$. Taking the trace in Eq. (4.1) in coordinate space, like in Eq. (1.28), immediately shows that $\hat{\rho}^{(1)}(t)$ is hermitian [2, 48, 49] and hence diagonalizable. The eigenvectors $\{|\Phi_k\rangle, k \in \mathbb{N}\}$ of $\hat{\rho}^{(1)}(t)$ yield an orthonormal basis for one particle in the interacting system,

$$\hat{\rho}^{(1)}(t)|\Phi_k(t)\rangle = f_k(t)|\Phi_k(t)\rangle \,. \tag{4.2}$$

The eigenvalues of Eq. (4.2) denote average occupation numbers $f_k(t)$ of the single particle states $\{|\Phi_k\rangle\}$ in the gas of $N$ interacting atoms.

In order to determine a condensate mode, we consider macroscopic occupation of only one mode in the long time limit $t \to \infty$:

$$\langle N_0 \rangle(\infty) = \sup\{f_k(\infty), k \in \mathbb{N}\} \sim \mathcal{O}(N) \,, \tag{4.3}$$

meaning that $\sigma_0(\infty) \equiv \langle N_0 \rangle(\infty)/N = $ const. in the thermodynamic limit, whereas all other modes remain weakly occupied, $f_k(\infty) \sim \mathcal{O}(1)$, for $k \neq 0$. Bose-Einstein condensation is now supposed to occur into the single particle mode $|\Phi_0\rangle \equiv |\Phi_0(\infty)\rangle$.

The replacement of the time dependent condensate mode by the equilibrium one becomes quantitatively accurate in either one of the following cases: For weak interactions $\xi \ll 1$, where the condensate state is approximately the ground state of the external trapping potential $|\Phi_0(t)\rangle \approx |\chi_0\rangle$ at all times, or, for initial states close to equilibrium, such that the condensate ket vector $|\Phi_0\rangle$ does not significantly change in time.

In order to quantify the condensate mode, we adopt [48, 49, 82] that, for dilute atomic gases at temperatures below $T_c$, $|\Phi_0\rangle$ is determined to terms $\mathcal{O}(N^{-1})$ by the Gross-Pitaevskii equation with all $N$ particles occupying the condensate mode:

$$\left[\frac{-\hbar^2 \vec{\nabla}^2}{2m} + V_{\text{ext}}(\vec{r}) + gN|\Psi_0(\vec{r})|^2 - \mu_0\right] \Psi_0(\vec{r}) = 0 \, , \tag{4.4}$$

For this reason, the approximate wave function $|\Psi_0\rangle$ in Eq. (4.4) is employed from now on, instead of the exact, but quantitatively unknown condensate mode $|\Phi_0\rangle$.

## 4.2 Interactions between condensate and non-condensate

In Section 4.1, we have defined the condensate mode $|\Psi_0\rangle$ in an interacting Bose gas. Here, a decomposition procedure for the full two body Hamiltonian in Eq. (3.4) is proposed, particularly for the nonlinear interaction term. The latter separates into three different types of two body interaction processes between condensate and non-condensate fields: single particle, pair and scattering events, which will be taken into account in the quantum master equation.

### 4.2.1 Separation of the second quantized field

The total bosonic field $\hat{\Psi}$ is expanded in the basis[1] $\{|\Psi_k\rangle, k \in \mathbb{N}_0\}$, where $|\Psi_0\rangle$ is the Gross-Pitaevskii ket in Eq. (4.4), and $\{|\Psi_k\rangle, k \in \mathbb{N}\}$ an arbitrary orthonormal, complete basis in the subspace of non-condensate particle wave functions. It thus separates into

---
[1] The choice of the basis states $\{|\Psi_k\rangle, k \in \mathbb{N}\}$ is arbitrary at this point, despite that the basis states have to be chosen pairwise orthogonal to $|\Psi_0\rangle$. Later in Chapter 4.3, we will choose the basis states $\{|\Psi_k\rangle, k \in \mathbb{N}\}$ such that they diagonalize the linearized non-condensate Hamiltonian.

$$\hat{\Psi} = |\Psi_0\rangle \hat{a}_0 + \sum_{k \neq 0} |\Psi_k\rangle \hat{a}_k \equiv \hat{\Psi}_0 + \hat{\Psi}_\perp \ . \tag{4.5}$$

In Eq. (4.5), $\hat{a}_k$ and $\hat{a}_k^\dagger$ are creation and annihilation operators, respectively, which satisfy usual bosonic commutation relations $\left[\hat{a}_k, \hat{a}_l^\dagger\right] = \delta_{kl}$ and $[\hat{a}_k, \hat{a}_l] = \left[\hat{a}_k^\dagger, \hat{a}_l^\dagger\right] = 0$. The corresponding Fock states on which these operators act are denoted by $|N_0, \{N_k\}\rangle$; the interpretation of a many particle Fock state is hence to find $N_0$ particles in the condensate mode $|\Psi_0\rangle$, and $\{N_k\} = \{N_1, N_2, \ldots\}$ in the non-condensate single particle modes $\{|\Psi_1\rangle, |\Psi_2\rangle, \ldots\}$.

### 4.2.2 Decomposition of the Hamiltonian

The following decomposition of the Hamiltonian only requires the validity of the Gross-Pitaevskii equation for the condensate field, $\hat{\Psi}_0(\vec{r})$, and the orthogonality of the two fields $\hat{\Psi}_0^\dagger$ and $\hat{\Psi}_\perp$, meaning that

$$\int_{\mathscr{C}} d\vec{r} \ \hat{\Psi}_0^\dagger(\vec{r}) \hat{\Psi}_\perp(\vec{r}) = 0 \ . \tag{4.6}$$

The Hamiltonian $\hat{\mathcal{H}}$ in second quantization, including two body interactions, is given by Eq. (3.4). The decomposition in Eq. (4.5) splits the Hamiltonian $\hat{\mathcal{H}}$ into three basic contributions

$$\hat{\mathcal{H}} = \hat{\mathcal{H}}_0 + \hat{\mathcal{H}}_\perp + \hat{V}_{0\perp} \ , \tag{4.7}$$

where $\hat{\mathcal{H}}_0$ and $\hat{\mathcal{H}}_\perp$ describe a pure condensate and non-condensate, respectively.

The condensate Hamiltonian $\hat{\mathcal{H}}_0$ contains the single particle contribution linear in the field $\hat{\Psi}_0$, as well as the nonlinear, self-interacting two body interaction term, and is given by

## 4.2. INTERACTIONS BETWEEN CONDENSATE AND NON-CONDENSATE

$$\hat{\mathcal{H}}_0 = \int_{\mathscr{C}} d\vec{r}\, \hat{\Psi}_0^\dagger(\vec{r}) \left[ -\frac{\hbar^2 \vec{\nabla}^2}{2m} + V_{\text{ext}}(\vec{r}) \right] \hat{\Psi}_0(\vec{r}) + \frac{g}{2} \int_{\mathscr{C}} d\vec{r}\, \hat{\Psi}_0^\dagger(\vec{r}) \hat{\Psi}_0^\dagger(\vec{r}) \hat{\Psi}_0(\vec{r}) \hat{\Psi}_0(\vec{r}) \ . \quad (4.8)$$

The Hamiltonian of the background gas $\hat{\mathcal{H}}_\perp$ includes only non-condensate field modes:

$$\hat{\mathcal{H}}_\perp = \int_{\mathscr{C}} d\vec{r}\, \hat{\Psi}_\perp^\dagger(\vec{r}) \left[ -\frac{\hbar^2 \vec{\nabla}^2}{2m} + V_{\text{ext}}(\vec{r}) \right] \hat{\Psi}_\perp(\vec{r}) + \frac{g}{2} \int_{\mathscr{C}} d\vec{r}\, \hat{\Psi}_\perp^\dagger(\vec{r}) \hat{\Psi}_\perp^\dagger(\vec{r}) \hat{\Psi}_\perp(\vec{r}) \hat{\Psi}_\perp(\vec{r}) \ . \quad (4.9)$$

The decomposition of the field in Eq. (4.5) furthermore induces different interaction terms *between* condensate and non-condensate fields, as evident from the Hamiltonian in Eq. (3.4), which are summarized by the term $\hat{\mathcal{V}}_{0\perp}$. It includes all possible two body interaction processes which will be separated and specified in the following section.

### 4.2.3 Two body interaction processes

The term $\hat{\mathcal{V}}_{0\perp}$ includes all possible two body interactions which can naturally be decomposed into three distinct kinds, according to the different net exchange of condensate particles, $\Delta N_0$, per interaction process. We call these different interaction events single particle ($\Delta N_0 = -\Delta N_\perp = \pm 1$, labeled by $\mathcal{X} = \rightsquigarrow$), pair ($\Delta N_0 = -\Delta N_\perp = \pm 2$, labeled by $\mathcal{X} = \rightsquigarrow\rightsquigarrow$) and scattering ($\Delta N_0 = \Delta N_\perp = 0$, labeled by $\mathcal{X} = \circlearrowright$) processes. Moreover, we distinguish condensate feeding and loss processes, corresponding to an effective annihilation ($\mathcal{X}$), or creation ($\mathcal{X}^*$) of a condensate atom.

In the following, it will be verified that all two body interactions can be split and identified as

$$\hat{\mathcal{V}}_{\perp 0} = \hat{\mathcal{V}}_{\rightsquigarrow} + \hat{\mathcal{V}}_{\rightsquigarrow\rightsquigarrow} + \hat{\mathcal{V}}_{\circlearrowright} \ . \quad (4.10)$$

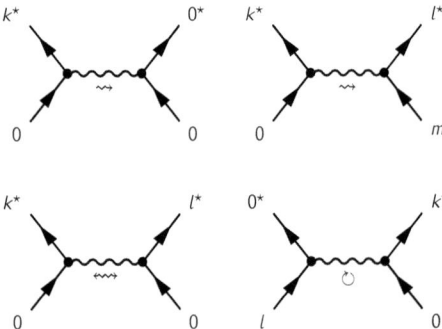

Figure 4.1: *Diagrammatic representation of all microscopic two body loss processes in Eq. (4.10). Upper two diagrams represent single particle losses (⤳), where one non-condensate particle is effectively created and one condensate atom is annihilated. The collection of all first order terms in the non-condensed field, $\mathcal{O}(\hat{\Psi}_\perp)$ (upper left diagrams), vanish in combination with crossed single particle terms, as a consequence of the Gross-Pitaevskii Eq. (4.4) and the orthogonality condition in Eq. (4.6). Lower diagrams display pair losses (⬳, lower left) and scattering processes (↻, lower right). Conjugate processes (not shown), related to condensate feeding, are obtained by exchanging the corresponding labels with respect to the diagram center.*

The different interaction terms are illustrated by a diagrammatic representation of the Hamiltonian's matrix elements [20, 45], meaning that a diagram

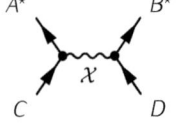

$$\equiv \frac{g}{Z}\zeta^{AB}_{CD}\hat{a}^\dagger_A\hat{a}^\dagger_B\hat{a}_C\hat{a}_D \qquad (4.11)$$

## 4.2. INTERACTIONS BETWEEN CONDENSATE AND NON-CONDENSATE

refers to an interaction process in which two particles $C$ and $D$ populating the single particle modes $\Psi_C$ and $\Psi_D$ are annihilated, resulting in a creation of two particles, $A$ and $B$, occupying subsequently the single particle modes $\Psi_A$ and $\Psi_B$. Each diagram is associated with a probability amplitude,

$$\zeta_{CD}^{AB} = \int_{\mathscr{C}} d\vec{r}\; \Psi_A^\star(\vec{r})\Psi_B^\star(\vec{r})\Psi_C(\vec{r})\Psi_D(\vec{r}) \;, \tag{4.12}$$

multiplied with the interaction strength $g/Z$, which gives the interaction energy of the different two body vertices, where $Z = 1$ for single particle ($\mathcal{X} = \leftrightsquigarrow$), $Z = 2$ for pair ($\mathcal{X} = \leftrightsquigarrow$) and $Z = 1/2$ for scattering ($\mathcal{X} = \circlearrowleft$) processes. The factor $Z$ is hence related to the multiple occurrences of condensate and non-condensate fields in the interaction term $\hat{V}_{0\perp}$ in Eq. (4.10).

The different types of two body diagrams — only including loss events, whereas feeding processes are formally obtained by exchanging the (time) arrows in the diagram — are depicted in Fig. 4.1: The nonlinear part of first order diagrams ($\mathcal{O}(\hat{\Psi}_\perp)$, upper left in Fig. 4.1) cancel out with mixed, single particle contributions between condensate and non-condensate fields in the Hamiltonian in Eq. (3.4). This is a consequence of the orthogonality of the two fields $\hat{\Psi}_0^\dagger(\vec{r})$ and $\hat{\Psi}_\perp(\vec{r})$, and of the fact that $\Psi_0(\vec{r})$ is the solution of the Gross-Pitaevskii equation in Eq. (4.4). Indeed, combining the upper left diagrams in Fig. 4.1 and their hermitian conjugates with mixed single particle contributions in Eq. (3.4), entails the term

$$\int_{\mathscr{C}} d\vec{r}\; \hat{\Psi}_\perp^\dagger(\vec{r}) \left[ \frac{-\hbar^2 \vec{\nabla}^2}{2m} + V_{\text{ext}}(\vec{r}) + \hat{\Psi}_0^\dagger(\vec{r})\hat{\Psi}_0(\vec{r}) \right] \hat{\Psi}_0(\vec{r}) + \text{h.c.} \simeq \mu_0 \int_{\mathscr{C}} d\vec{r}\; \hat{\Psi}_\perp^\dagger(\vec{r})\hat{\Psi}_0(\vec{r}) + \text{h.c.} = 0 \;, \tag{4.13}$$

which vanishes for sufficiently weak interactions,[2] because of Eqs. (4.4, 4.6).

All remaining diagrams contribute to interactions between condensate and non-

---
[2] since in this case, $\Psi_0(\vec{r})$ is still an approximate solution of the Gross-Pitaevskii equation (4.4) for arbitrary $N_0$ with eigenvalue $\mu_0$

condensate, and are grouped into single particle (according to $\Delta N_0 = \pm 1$ and $\Delta N_\perp = \mp 1$ - upper right diagram in Fig. 4.1), pair ($\Delta N_0 = \pm 2$ and $\Delta N_\perp = \mp 2$ - lower left diagram in Fig. 4.1) and scattering ($\Delta N_0 = 0$ and $\Delta N_\perp = 0$ - lower right diagram in Fig. 4.1) events.

Generally, a pair event ($\mathcal{O}(\hat{\Psi}_\perp^2)$, lower left diagram in Fig. 4.1) effectively creates (annihilates) two condensate particles and annihilates (creates) two non-condensate particles. Pair loss events (see bottom - left diagram) destroy two condensate and create two non-condensate particles by self-interaction in the condensate. Vice versa, a pair feeding event is an interaction of two non-condensed atoms, creating two condensate particles (bottom - right diagram). These processes can consequently be summarized by all diagrams of the type

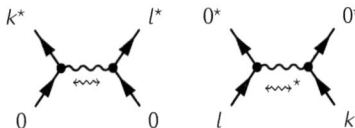

which represent the interaction term

$$\hat{V}_{\rightsquigarrow} = g \int_\mathscr{C} d\vec{r}\, \hat{\Psi}_\perp^\dagger(\vec{r}) \hat{\Psi}_\perp^\dagger(\vec{r}) \hat{\Psi}_0(\vec{r}) \hat{\Psi}_0(\vec{r}) + g \int_\mathscr{C} d\vec{r}\, \hat{\Psi}_0^\dagger(\vec{r}) \hat{\Psi}_0^\dagger(\vec{r}) \hat{\Psi}_\perp(\vec{r}) \hat{\Psi}_\perp(\vec{r}) \ . \qquad (4.14)$$

Scattering events ($\mathcal{O}(\hat{\Psi}_\perp^2)$, lower right diagram in Fig. 4.1) leave the condensate and non-condensate particle number unchanged, $\Delta N_0 = \Delta N_\perp = 0$:

$$\hat{V}_\circlearrowleft = 2g \int_\mathscr{C} d\vec{r}\, \hat{\Psi}_\perp^\dagger(\vec{r}) \hat{\Psi}_0^\dagger(\vec{r}) \hat{\Psi}_\perp(\vec{r}) \hat{\Psi}_0(\vec{r}) \ . \qquad (4.15)$$

Finally, the last and most important type of two body contributions during the

## 4.3. HAMILTONIAN OF THE NON-CONDENSATE BACKGROUND GAS

process of Bose-Einstein condensation are single particle events ($\mathcal{O}(\hat{\Psi}_\perp^3)$, upper right in Fig. 4.1). Single particle losses originate from interactions between one non-condensate and one condensate particle, which lead to an annihilation of the condensate and creation of one non-condensate particle (bottom – left diagram). Here, a single particle feeding process is a creation of a condensate particle, which originates from self-interaction in the background gas (bottom – right diagram). The sum of all these events are all diagrams of the type

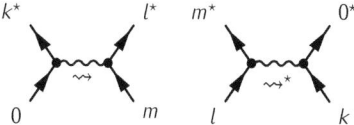

which correspond to the interaction term

$$\hat{V}_{\rightsquigarrow} = \frac{g}{2} \int_{\mathscr{C}} d\vec{r}\, \hat{\Psi}_\perp^\dagger(\vec{r}) \hat{\Psi}_\perp^\dagger(\vec{r}) \hat{\Psi}_\perp(\vec{r}) \hat{\Psi}_0(\vec{r}) + \frac{g}{2} \int_{\mathscr{C}} d\vec{r}\, \hat{\Psi}_0^\dagger(\vec{r}) \hat{\Psi}_\perp^\dagger(\vec{r}) \hat{\Psi}_\perp(\vec{r}) \hat{\Psi}_\perp(\vec{r}) \,. \qquad (4.16)$$

We subsequently get the exact decomposition of the total Hamiltonian $\hat{\mathcal{H}}$ into

$$\hat{\mathcal{H}} = \hat{\mathcal{H}}_0 + \hat{\mathcal{H}}_\perp + \hat{V}_{\rightsquigarrow} + \hat{V}_{\rightsquigarrow\rightsquigarrow} + \hat{V}_{\circlearrowright} \,, \qquad (4.17)$$

where the different interaction terms, $\hat{V}_{\rightsquigarrow}$ in Eq. (4.16), $\hat{V}_{\rightsquigarrow\rightsquigarrow}$ in Eq. (4.14) and $\hat{V}_{\circlearrowright}$ in Eq. (4.15), describe single particle ($\rightsquigarrow$), pair ($\rightsquigarrow\rightsquigarrow$) and scattering ($\circlearrowright$) processes between condensate and non-condensate atoms.

## 4.3 Hamiltonian of the non-condensate background gas

The non-condensed cloud is a rapidly decorrelating and thermalizing gas of particles. We describe this thermalization process by the coupling of the non-condensate to a heat reservoir, thus using the $N$-body Born-Markov ansatz for the state of the gas (see Chapter 3), adding the irreversible decay of non-condensate correlations arising from the thermalization process [59, 63, 64] by the transformation:

$$\left\langle \hat{\mathscr{E}}^\dagger(\vec{r},\tau)\hat{\mathscr{E}}(\vec{r}\,',0) \right\rangle \to \exp[-\Gamma^2\tau^2]\left\langle \hat{\mathscr{E}}^\dagger(\vec{r},\tau)\hat{\mathscr{E}}(\vec{r}\,',0) \right\rangle_{\text{Born-Markov}}, \quad (4.18)$$

with $\Gamma = \tau_{\text{col}}^{-1}$ the thermalization rate, $\hat{\mathscr{E}}, \hat{\mathscr{E}}^\dagger : \mathscr{F}_\perp \to \mathscr{F}_\perp$ arbitrary non-condensate field operators (see Chapter 5), and where the right hand side of Eq. (4.18) indicates the ensemble average over the Bose gas' state $\hat{\sigma}^{(N)}(t)$ in Eq. (3.8) in $N$-body Born-Markov approximation. Omitting negligible $\mathcal{O}(g^3)$-contributions (arising from the nonlinear self interaction term of the non-condensate Hamiltonian $\hat{\mathcal{H}}_\perp$) from the master equation describing the dynamics of two body collisions ($\sim g^2$), only the linearized part of the non-condensate Hamiltonian $\hat{\mathcal{H}}_\perp$ is needed to describe the thermalized non-condensate state occuring in the correlation functions of the type like in Eq. (4.18). The non-condensate Hamiltonian is thus replaced according to

$$\hat{\mathcal{H}}_\perp \simeq \int_{\mathscr{C}} d\vec{r}\, \hat{\Psi}_\perp^\dagger(\vec{r}) \left[ \frac{-\hbar^2 \vec{\nabla}^2}{2m} + V_{\text{ext}}(\vec{r}) \right] \hat{\Psi}_\perp(\vec{r}) + \mathcal{O}(g). \quad (4.19)$$

We will see in Chapter 5 in more detail that the thermalization process occuring on a time scale $\tau_{\text{col}}$ justifies the Born-Markov dynamics assumption for a dilute Bose-Einstein condensate [59, 63, 64].

### 4.3.1 Diagonalization of the non-condensate Hamiltonian

So far, the non-condensate single particle wave functions $|\Psi_k\rangle$ have been chosen arbitrarily (pairwise orthogonal to $|\Psi_0\rangle$). They are now quantified by the constraint of diagonalizing the non-condensate Hamiltonian in Eq. (4.19).

To this end, we introduce arbitrary unitary matrices $\mathcal{T}$ and $\mathcal{T}^\dagger = \mathcal{T}^{-1}$, and an arbitrary orthonormal complete basis[3] $\{|\Theta_k\rangle, k \in \mathbb{N}\}$ spanning the single par-

---
[3] The basis $\{|\Theta_k\rangle, k \in \mathbb{N}\}$ can always be constructed out of the single particle states $|\chi_k\rangle$ of the non-interacting system by the Gram-Schmidt procedure, which, however, does

## 4.3. HAMILTONIAN OF THE NON-CONDENSATE BACKGROUND GAS

ticle subspace orthogonal to $|\Psi_0\rangle$. Let's denote the corresponding bosonic creation/annihilation operators by $\hat{\gamma}_k$ and $\hat{\gamma}_k^\dagger$. Expanding the total bosonic field according to Eq. (4.5) in the basis $|\Theta_k\rangle$, and transforming the particle operators $\hat{\gamma}_k$ and $\hat{\gamma}_k^\dagger$ according to

$$\hat{\gamma}_k = \sum_{l \neq 0} \mathcal{T}_{kl} \hat{a}_l \quad \text{and} \quad \hat{\gamma}_k^\dagger = \sum_{l \neq 0} \mathcal{T}_{kl}^* \hat{a}_l^\dagger , \quad (4.20)$$

respectively, for $k \neq 0$, leads to the representation of the second quantized field in a new basis set:

$$\hat{\Psi} = \hat{\Psi}_0 + \sum_{l \neq 0} |\Psi_l\rangle \hat{a}_l , \quad (4.21)$$

where the states $|\Psi_l\rangle$ are superpositions of the single particle states $|\Theta_k\rangle$:

$$|\Psi_l\rangle = \sum_{k \neq 0} \mathcal{T}_{kl} |\Theta_k\rangle . \quad (4.22)$$

The basis $\{|\Psi_k\rangle, k \in \mathbb{N}\}$ has the same properties as $\{|\Theta_k\rangle, k \in \mathbb{N}\}$: It is orthonormal, $\langle \Psi_k | \Psi_l \rangle = \delta_{kl}$, it is complete in the subspace of non-condensate single particle wave functions,

$$\sum_{k \neq 0} |\Psi_k\rangle\langle\Psi_k| = \sum_{k \neq 0} |\Theta_k\rangle\langle\Theta_k| = \hat{1} - |\Psi_0\rangle\langle\Psi_0| , \quad (4.23)$$

as well as each $|\Psi_k\rangle$ is orthogonal to $|\Psi_0\rangle$. Since $\mathcal{T}$ is just a unitary transformation, it is easy to verify that the operators $\hat{a}_k^\dagger$ and $\hat{a}_k$ satisfy usual bosonic commutations relations:

---

not necessarily diagonalize the non-condensate Hamiltonian in Eq. (4.19). Since each ket $|\Theta_k\rangle$ is orthogonal to $|\Psi_0\rangle$, it is evident that $|\Theta_k\rangle \neq |\chi_k\rangle$ in general.

$$\left[\hat{a}_k, \hat{a}_l^\dagger\right] = \sum_{j,j'\neq 0}\left[\hat{\gamma}_j, \hat{\gamma}_{j'}^\dagger\right]T^*_{kj}T_{lj'} = \sum_{j\neq 0}T^*_{kj}T_{lj} = \delta_{kl} \ . \qquad (4.24)$$

The basis change in Eqs. (4.20, 4.22) conserves the total number of particles due to the invariance of the number operator $\hat{N} = \int_{\mathscr{C}} d\vec{r}\, \Psi^\dagger(\vec{r})\hat{\Psi}(\vec{r})$ under the transformation $T$. By introducing infinite dimensional vectors $\vec{a} = (\hat{a}_1, \hat{a}_2, ...)$ and $\vec{\gamma} = (\hat{\gamma}_1, \hat{\gamma}_2, ...)$, Eq. (4.20) turns into

$$\vec{\gamma} = T\vec{a} \quad \text{and} \quad \vec{\gamma}^\dagger = \vec{a}^\dagger T^\dagger \ . \qquad (4.25)$$

To show that the basis $\{|\Psi_k\rangle, k \in \mathbb{N}\}$ diagonalizes $\hat{\mathcal{H}}_\perp$ for the correct choice of the matrices $T$ and $T^\dagger$, the Hamiltonian in Eq. (4.19) is rewritten into a general bilinear form

$$\hat{\mathcal{H}}_\perp = \vec{a}^\dagger T^\dagger \epsilon T \vec{a} \ , \qquad (4.26)$$

defining the non-diagonal energy tensor $\epsilon$:

$$\epsilon_{kk'} = \sum_l \langle\Theta_k|\chi_l\rangle\langle\chi_l|\Theta_{k'}\rangle\eta_l \ , \qquad (4.27)$$

where $|\chi_k\rangle$ are the eigenstates of the first quantized Hamiltonian $h_1 = [\vec{p}^2/2m + V_{\text{ext}}(\vec{r})]$ for non-interacting particles, and $\eta_k$ the corresponding unperturbed eigenenergies.

Since the tensor $\epsilon$ is hermitian, $\epsilon$ itself and hence $\hat{\mathcal{H}}_\perp$ are diagonalizable by a unitary matrix $\mathcal{M}$, which is defined by the equation

$$\epsilon = \mathcal{M} \cdot \begin{bmatrix} \epsilon_1 & 0 & 0 \\ 0 & \epsilon_2 & 0 \\ 0 & 0 & \ddots \end{bmatrix} \cdot \mathcal{M}^\dagger \ . \qquad (4.28)$$

## 4.3. HAMILTONIAN OF THE NON-CONDENSATE BACKGROUND GAS

Here, the $\epsilon_k$ mark the single particle energies of non-condensate single particle states $|\Psi_k\rangle$.

The relation between the matrices $\mathcal{M}$ and $\mathcal{T}$ is now obvious from Eq. (4.26), i.e., the choice $\mathcal{T} = \mathcal{M}^\dagger$ diagonalizes $\hat{\mathcal{H}}_\perp$ in the basis $\{|\Psi_k\rangle, k \in \mathbb{N}\}$:

$$\hat{\mathcal{H}}_\perp = \sum_{k \neq 0} \epsilon_k \hat{a}_k^\dagger \hat{a}_k = \sum_{k \neq 0} \epsilon_k \hat{N}_k \ . \tag{4.29}$$

Single particle states $|\Psi_k\rangle$ are hence interpreted as non-condensate particles,[4] being uniquely quantified by Eqs. (4.22, 4.27, 4.28).

### 4.3.2 Perturbative spectrum of non-condensate particles

For completeness, the relation between the perturbative expansion of the single particle states $|\Psi_k\rangle$ in terms of the small parameter $\xi = a\varrho^{1/3}$ and the perturbative spectrum of non-condensate particles in Eq. (4.29) is presented here. The condensate wave function $|\Psi_0\rangle$ in Eq. (4.4) can be considered as a function of the parameter $a\varrho = \xi\varrho^{2/3}$, given the fact that the atomic density $\varrho$ can be replaced by the peak density $N|\Psi_0(0)|^2$ at the center of the trap. Note that the dimensionless parameter $\xi = a\varrho^{1/3} \ll 1$ is the small parameter of our theory. It is thus possible to expand the Gross-Pitaevskii state $|\Psi_0\rangle$ around $\xi = 0$, the formal limit of unperturbed eigenstates:

$$|\Psi_0\rangle = |\chi_0\rangle + \sum_{m \neq 0} \xi^m |\Psi_0^{(m)}\rangle \ , \tag{4.30}$$

where the expansion coefficients $|\Psi_0^{(m)}\rangle$ are independent of $\xi$:

$$|\Psi_0^{(m)}\rangle = \frac{1}{m!} \frac{\partial^m |\Psi_0\rangle}{\partial^m \xi}\bigg|_{\xi=0} \ . \tag{4.31}$$

---
[4] The kets $|\Psi_k\rangle$ are of course in general *not* equal to the single particle states $|\chi_k\rangle$ of the non-interacting system for a finite interaction strength $g \neq 0$.

In Eq. (4.30), the unperturbed state $|\chi_0\rangle$ is the solution of the Schrödinger equation, obtained by formally setting $a = 0$ in Eq. (4.4). Non-condensate single particle basis states $|\Psi_k\rangle$ in the interacting system can obviously be expanded as well in an infinite series of the parameter $\xi$. They also turn into the eigenstates of a non-interacting gas $|\chi_k\rangle$ for all $k \in \mathbb{N}_0$ in the formal limiting case $\xi \to 0^+$ at constant density $\varrho$.

The physical interpretation of the above analysis is that interactions between the atoms exchange not only particles between the different single particle modes due to atomic scattering, but moreover perturb the shape of the particles' quantum mechanical wave functions. Expansion of the states $|\Psi_k\rangle$ in terms of the parameter $\xi$ like in Eq. (4.31) entails the following expansion of the energy tensor $\epsilon$ in Eq. (4.27):

$$\epsilon_{kl} = \eta_k \delta_{kl} + \sum_{m \neq 0} \xi^m \epsilon_{kl}^{(m)} \, . \qquad (4.32)$$

Hence, the expansion in Eq. (4.32) guides us to an energy tensor $\epsilon$, which is diagonal in the $0^{th}$ order in $\xi$, with unperturbed single particle energies $\eta_k$. Corrections scale as $\xi^k$ with weighting coefficients $\epsilon_{kk'}^{(m)}$ given by

$$\epsilon_{kk'}^{(m)} = \frac{1}{m!} \left\{ \langle \Psi_k^{(m)} | \chi_l \rangle \eta_l + \langle \Psi_l^{(m)} | \chi_k \rangle^* \eta_k + \sum_{\substack{l \\ 0 < z < m}} \left[ \langle \Psi_k^{(z)} | \chi_l \rangle \langle \Psi_{k'}^{(m-z)} | \chi_l \rangle^* \right] (1 - \delta_{m1}) \right\} \, . \qquad (4.33)$$

This yields the spectrum for non-condensate single particle states

$$\epsilon_k = \eta_k + \sum_{m \neq 0} \xi^m \left[ \mathcal{M}^\dagger \cdot \epsilon^{(m)} \cdot \mathcal{M} \right]_{kk} \, , \qquad (4.34)$$

reflecting that single particle energies $\epsilon_k$ of non-condensate states are shifted by the interaction in the Bose gas. Indeed, they turn into single particle energies $\eta_k$ of the non-interacting system, as well as $|\Psi_k\rangle \to |\chi_k\rangle$ in the formal perturbative limit

$\xi \to 0^+$.

## 4.4 Hilbert spaces

We finally analyze the underlying Hilbert space of single particle wave functions, $\mathcal{H}$, and the many particle Fock-Hilbert space $\mathcal{F}$ of the Bose gas. For the latter case, we distinguish the Fock-Hilbert space $\mathcal{F}(N)$ containing all Fock states of a fixed particle number $N$ from the general (extended) Fock-Hilbert space $\mathcal{F}$ spanned by the set of all possible linear combinations of multi-mode Fock states $|\{N_k\}\rangle$.

### 4.4.1 Single particle Hilbert space

An eigenbasis for one particle in the interacting system is defined by the single particle basis $\{|\Psi_k\rangle, k \in \mathbb{N}_0\}$, which spans the complete Hilbert space of single particle states, see Eq. (4.23). The Hilbert space of single particle wave functions is consequently defined by

$$\mathcal{H} = \text{span}\{|\Psi_0\rangle, |\Psi_1\rangle, |\Psi_2\rangle, \ldots\} \equiv \mathcal{H}_0 \oplus \mathcal{H}_\perp, \tag{4.35}$$

with $\mathcal{H}_0 = \text{span}\{|\Psi_0\rangle\}$, the space of condensate wave functions, and with the counterpart $\mathcal{H}_\perp = \text{span}\{|\Psi_k\rangle, k \in \mathbb{N}\}$, the space of non-condensate single particle wave functions, which are orthogonal to the condensate wave function $|\Psi_0\rangle$.

### 4.4.2 Fock-Hilbert space

Here, we aim at emphasizing the tensor structure of the underlying Fock-Hilbert space $\mathcal{F} = \mathcal{F}_0 \otimes \mathcal{F}_\perp$ with Fock basis elements $|\{N_k\}\rangle$, which applies even for finite particle numbers. The Fock number states refer to particle occupations of the underlying single particle wave functions $\{|\Psi_k\rangle, k \in \mathbb{N}\}$, defining the functional space $\mathcal{H}$ in Section 4.4.1. Our Fock-Hilbert space $\mathcal{F}$ is spanned by an infinite countable set of vectors [83, 84]

$$|N_0\rangle \otimes |\{N_k\}\rangle = |N_0, N_1, ...\rangle ,\qquad (4.36)$$

where $N_0$ and $\{N_k\} = N_1, N_2, ...$ are arbitrary sequences of integer numbers ($N_k = 0, 1, 2, ...$). In our treatment, the basis state $|N_0\rangle \otimes |\{N_k\}\rangle$ thus refers to a state with $N_0$ particles occupying the condensate mode $|\Psi_0\rangle$, and $\{N_k\}$ particles the non-condensate modes $\{|\Psi_k\rangle, k \in \mathbb{N}\}$. The total Fock-Hilbert space is thus simply a product space,

$$\mathscr{F} = \mathscr{F}_0 \otimes \mathscr{F}_\perp ,\qquad (4.37)$$

with the condensate Fock-Hilbert space $\mathscr{F}_0 = \mathrm{span}\{|N_0\rangle : N_0 \in \mathbb{N}\}$, and the non-condensate Fock-Hilbert space $\mathscr{F}_\perp = \mathrm{span}\{|N_1, N_2, ...\rangle : N_k \in \mathbb{N}\}$. In the following, the partial traces over the two subsystems condensate and non-condensate are thus taken with respect to the basis elements of the two Hilbert spaces $\mathscr{F}_0$ and $\mathscr{F}_\perp$ in Eq. (4.37).

### 4.4.3 Fock-Hilbert space of states with fixed particle number

All Fock states of a fixed particle number in the Bose gas of $N$ atoms are elements of a reduced Hilbert space, $\mathscr{F}(N)$, including the linear combination of all states $|N_0, \{N_k\}\rangle$ with $\sum_k N_k = N$. The latter can be related to the total Fock-Hilbert space $\mathscr{F}$ in Section 4.4.2. Consider a given state with $N_0$ particles in the condensate, and consequently $(N - N_0)$ particles in the non-condensate. The corresponding space $\mathscr{F}^{(N)}(N_0)$ is a subset of $\mathscr{F} = \mathscr{F}_0 \otimes \mathscr{F}_\perp$,

$$\mathscr{F}^{(N)}(N_0) = \mathrm{span}\{|N_0\rangle\} \otimes \mathscr{F}_\perp(N - N_0) \subset \mathscr{F}_0 \otimes \mathscr{F}_\perp ,\qquad (4.38)$$

## 4.4. HILBERT SPACES

where $\mathscr{F}_\perp(N-N_0)$ is the set of all possible non-condensate Fock states with $(N-N_0)$ particles:

$$\mathscr{F}_\perp(N-N_0) = \text{span}\left\{|\{N_k\}\rangle : \sum_{k\neq 0} N_k = (N-N_0)\right\}. \quad (4.39)$$

The entire Hilbert space $\mathscr{F}(N)$ of states with fixed particle number $N$ is the direct sum of the subspaces $\mathscr{F}(N_0)$:

$$\mathscr{F}(N) = \bigoplus_{N_0=0}^{N} \mathscr{F}^{(N)}(N_0) = \text{span}\left\{\bigoplus_{N_0=0}^{N} |N_0\rangle \otimes \mathscr{F}_\perp(N-N_0)\right\}. \quad (4.40)$$

Therefore, the reduced Fock-Hilbert space $\mathscr{F}(N)$ is in general different from the total Fock-Hilbert space $\mathscr{F}$ in Eq. (4.37).

The constraint of particle number conservation, however, need *not* be imposed onto the Fock spaces, but can be imposed onto the *state* of the system, thus simplifying and disentangling the two Fock-Hilbert spaces of condensate and non-condensate, as in Eq. (4.37). This way formally allows all occupations to vary from $N_k = 0 \ldots \infty$, and thus the total number of atoms $N$ to vary from $N = 0 \ldots \infty$ in the partial traces of a general operator average.

Particle number conservation is thus accounted for by defining the (in general mixed) $N$-body state of fixed particle number $N$ as a map $\hat{\sigma}^{(N)}(t) : \mathscr{F} \to \mathscr{F}(N)$ in order to ensure that any state $|N_0, \{N_k\}\rangle$ with $\sum_k N_k \neq N$, in a gas of $N$ atoms, has zero probability to occur and therefore does not contribute to an operator average taken over $\mathscr{F}$. The state $\hat{\sigma}^{(N)}(t)$ of the Bose gas is hence formally defined by the constraint that it maps any number state $|\Psi\rangle$ with total particle number different from $N$ to zero, i.e.

$$\hat{\sigma}^{(N)} : \mathscr{F} \to \mathscr{F}(N) \quad \text{with} \quad \text{Ker}\{\hat{\sigma}^{(N)}\} = \left\{|\{N_k\}\rangle : \sum_k N_k \neq N\right\}. \quad (4.41)$$

An example of such a state is given by the $N$-body Born ansatz, see Eq. (3.8), which we will use in the derivation of the master equation in Chapter 5.

# Chapter 5

# Lindblad master equation for a Bose-Einstein condensate

In this chapter, the Lindblad quantum master equation for the Bose-Einstein phase transition in a Bose gas of $N$ atoms is derived under the constraint of particle number conservation and within the Markovian dynamics assumption. This quantum master equation describes the time evolution of the condensate *and* non-condensate particle number distribution during the relaxation of the full N-body state $\hat{\sigma}^{(N)}(t)$ of the gas to the Bose-condensed phase. We give analytical expressions for the transition rates and energy shifts corresponding to the various two particle interaction processes specified in Chapter 4.

## 5.1 Evolution equation of the total density matrix

In analogy to the standard quantum optical derivation [20, 21, 77], we start with the von-Neumann equation for the many particle state $\hat{\sigma}^{(N)}(t): \mathscr{F} \longmapsto \mathscr{F}(N)$ of fixed particle number $N$ (see Section 4.4.3), defined on the Fock-Hilbert space $\mathscr{F} = \mathscr{F}_0 \otimes \mathscr{F}_\perp$ in Eq. (4.37):

$$\frac{\partial \hat{\sigma}^{(N)}(t)}{\partial t} = -\frac{i}{\hbar}\left[\hat{\mathcal{H}}, \hat{\sigma}^{(N)}(t)\right] , \qquad (5.1)$$

where $\hat{\mathcal{H}}$ is the total many particle Hamiltonian including two body interactions in Eq. (3.4). Since the N-body state includes only Fock states of fixed particle number according to its definition in Eq. (4.41), it commutes initially with the total number of atoms, $\left[\hat{\sigma}^{(N)}(0), \hat{N}\right] = 0$. As the number of atoms is conserved during the further time evolution, $\left[\hat{\mathcal{H}}, \hat{N}\right] = 0$, the state doesn't change its particle number, so that

$$\left[\hat{\sigma}^{(N)}(t), \hat{N}\right] = 0 \tag{5.2}$$

for any time. The Hamiltonian $\hat{\mathcal{H}}$ has been shown to split in the standard (quantum optical) fashion into

$$\hat{\mathcal{H}} = \hat{\mathcal{H}}_0 + \hat{\mathcal{H}}_\perp + \hat{V}_{0\perp} , \tag{5.3}$$

where $\hat{\mathcal{H}}_0$ is the condensate Hamiltonian in Eq. (4.8), $\hat{\mathcal{H}}_\perp$ the non-condensate Hamiltonian in Eq. (4.19), and $\hat{V}_{0\perp}$ the two body interaction term in Eq. (4.10). As discussed in Section 4.3, the thermalization process due to interactions within the non-condensate is modeled by coupling the non-condensate to a thermal bath (see Section 5.2.1).

With the decomposition of $\hat{\mathcal{H}}$ in Eq. (5.3), the von-Neumann equation turns into

$$\frac{\partial \hat{\sigma}^{(N)}(t)}{\partial t} = -\frac{i}{\hbar}\left[\hat{\mathcal{H}}_0, \hat{\sigma}^{(N)}(t)\right] - \frac{i}{\hbar}\left[\hat{\mathcal{H}}_\perp, \hat{\sigma}^{(N)}(t)\right] - \frac{i}{\hbar}\left[\hat{V}_{0\perp}, \hat{\sigma}^{(N)}(t)\right] . \tag{5.4}$$

Now, all operators, i.e., the condensate and the non-condensate field, $\hat{\Psi}_0(\vec{r})$ and $\hat{\Psi}_\perp(\vec{r})$, as well as the density matrix $\hat{\sigma}^{(N)}(t)$ are transformed into the interaction picture (denoted by the label $I$) with respect to the Hamiltonian parts $\hat{\mathcal{H}}_0$ and $\hat{\mathcal{H}}_\perp$. Consequently, the different operators undergo the following transformation (consult Appendix A.1 for a detailed evaluation of $\hat{\Psi}_0^{(I)}(\vec{r}, t)$ and $\hat{\Psi}_\perp^{(I)}(\vec{r}, t)$):

## 5.1. EVOLUTION EQUATION OF THE TOTAL DENSITY MATRIX

$$\hat{x}(t) \to \hat{x}^{(I)}(t) = \hat{\mathcal{U}}(t)\hat{x}\hat{\mathcal{U}}^\dagger(t) ,\tag{5.5}$$

where the time evolution operator $\hat{\mathcal{U}}(t)$ is given by

$$\hat{\mathcal{U}}(t) = \exp\left[\frac{i}{\hbar}\left(\hat{\mathcal{H}}_0 + \hat{\mathcal{H}}_\perp\right)t\right] = \exp\left[\frac{i}{\hbar}\hat{\mathcal{H}}_0 t\right]\exp\left[\frac{i}{\hbar}\hat{\mathcal{H}}_\perp t\right] \equiv \hat{\mathcal{U}}_0(t)\hat{\mathcal{U}}_\perp(t) ,\tag{5.6}$$

since $\left[\hat{\mathcal{H}}_0, \hat{\mathcal{H}}_\perp\right] \equiv 0$. The time evolution operator $\hat{\mathcal{U}}(t)$ leads to $\hat{a}_0(t) = \hat{a}_0 e^{-i\mu_0 t/\hbar}$ and $\hat{a}_k(t) = \hat{a}_k e^{-i\epsilon_k t/\hbar}$ (see Appendix A.1), leading to the time dependent condensate and non-condensate fields in the interaction picture, $\hat{\Psi}_0(\vec{r}) \to \hat{\Psi}_0^{(I)}(\vec{r},t) = \hat{a}_0 e^{-i\mu_0 t/\hbar}\Psi_0(\vec{r})$ and $\hat{\Psi}_\perp(\vec{r}) \to \hat{\Psi}_\perp^{(I)}(\vec{r},t) = \sum_{k\neq 0} \hat{a}_k e^{-i\epsilon_k t/\hbar}\Psi_k(\vec{r})$.

The time evolution of the full density operator $\hat{\sigma}^{(N,I)}(t)$ in the interaction picture is consequently determined by the interaction between condensate and non-condensate particles, according to:

$$\frac{\partial \hat{\sigma}^{(N,I)}(t)}{\partial t} = -\frac{i}{\hbar}[\hat{\mathcal{V}}_{0\perp}^{(I)}(t), \hat{\sigma}^{(N,I)}(t)] .\tag{5.7}$$

Considering sufficiently dilute Bose gases, $a\varrho^{1/3} \ll 1$, and thus only two body interactions, Eq. (5.7) can be solved in the second order[1] iteration [20] in $\hat{\mathcal{V}}_{0\perp}^{(I)}(t)$. To this end, Eq. (5.7) is integrated between the time interval $t$ and $t + \Delta t$, where $\tau_{\text{col}} < \Delta t \ll \tau_0$ according to the separation of time scales[2] in a Bose gas:

$$\hat{\sigma}^{(N,I)}(t + \Delta t) = \hat{\sigma}^{(N,I)}(t) - \frac{i}{\hbar}\int_t^{t+\Delta t} dt' \left[\hat{\mathcal{V}}_{0\perp}^{(I)}(t'), \hat{\sigma}^{(N,I)}(t')\right] .\tag{5.8}$$

---

[1] This second order iteration is necessary, because the first order iteration vanishes (see Section 5.3).
[2] Remember that, in the present case, the time scale for $\Delta t$ is to be chosen slightly larger than the average time of two body collisions according to the separation of time scales in a Bose-Einstein condensate (see Section 3.1).

Iteration up to second order in the interaction term $\hat{\mathcal{V}}_{0\perp}^{(l)}(t)$ – containing interaction terms up to order $g^2$ – finally leads to

$$\Delta\hat{\sigma}^{(N,l)}(t) = -\frac{i}{\hbar}\int_{t}^{t+\Delta t} dt' \left[\hat{\mathcal{V}}_{0\perp}^{(l)}(t'), \hat{\sigma}^{(N,l)}(t)\right] - \int_{t}^{t+\Delta t} dt' \int_{t}^{t'} \frac{dt''}{\hbar^2} \left[\hat{\mathcal{V}}_{0\perp}^{(l)}(t'), \left[\hat{\mathcal{V}}_{0\perp}^{(l)}(t''), \hat{\sigma}^{(N,l)}(t)\right]\right], \quad (5.9)$$

where $\Delta\hat{\sigma}^{(N,l)}(t) = \hat{\sigma}^{(N,l)}(t+\Delta t) - \hat{\sigma}^{(N,l)}(t)$. Note that the time integral over $t''$ is replaced by $t$, setting $\hat{\sigma}^{(N,l)}(t'') \to \hat{\sigma}^{(N,l)}(t)$ on the right hand side of Eq. (5.9), which corresponds to the Markov dynamics assumption (see below). Breaking the iteration procedure in $\hat{\mathcal{V}}_{0\perp}$ in the second order suffices to model the dynamics of two body interactions in dilute atomic gases, $a\varrho^{1/3} \ll 1$, since they are described by the terms proportional to $g^2$ within the time scale $\Delta t$ of the temporal iteration of Eq. (5.9).

## 5.2 Time evolution of the reduced condensate density matrix

In order to access the time evolution of the reduced condensate subsystem in the presence of the non-condensate gas, the partial trace over $\mathcal{F}_\perp$ in Eq. (5.9) needs to be taken. Here, we explain the N-body Born ansatz which leads to a time local quantum master equation for the reduced condensate density matrix.

### 5.2.1 N-body Born ansatz

The standard quantum optical ansatz [17, 18, 19, 20, 21] to derive a Markov quantum master equation of Lindblad type considers the non-condensate as a large (undepleted) thermal bath and separates its dynamics from the condensate subsystem. In the present case, however, this is not possible because the non-condensate exchanges particles with the condensate, so that

$$\hat{\sigma}^{(N)}(t) \neq \hat{\rho}_0(t) \otimes \hat{\rho}(T) . \quad (5.10)$$

## 5.2. TIME EVOLUTION OF THE REDUCED CONDENSATE DENSITY MATRIX

The circumvention of this ansatz is conceptually essential for the dynamics of a Bose gas of exactly $N$ particles. However, the standard Born ansatz in Eq. (5.10) can be generalized by describing the non-condensate as a series of thermal states with different particle numbers $(N - N_0)$, given that $N_0$ particles occupy the condensate mode. This ansatz is physically justified, since the non-condensate thermalizes rapidly [51, 64], therefore decohering all off-diagonal elements between subspaces of different non-condensate particle numbers.

This rapid non-condensate thermalization is formally taken into account by the map $\mathscr{D}$ obeying the following properties: (i) It does not change the particle number in the gas, (ii) it erases coherences between states of different particle numbers $(N - N_0)$ and $(N - M_0)$ in the non-condensate, and (iii) it turns each non-condensate state of $(N - N_0)$ particles into a thermal state of corresponding particle number occupation. From (i)-(iii), it follows that the $N$-body state of the Bose gas is diagonal in particle number representation:

$$\hat{\sigma}^{(N)}(t) \approx \mathscr{D}\left(\hat{\sigma}^{(N)}(t)\right) \equiv \sum_{N_0=0}^{N} p_N(N_0, 0) |N_0\rangle\langle N_0| \otimes \hat{\rho}_\perp(N - N_0, T) , \qquad (5.11)$$

where each condensate state of $N_0$ particles necessarily implies a condensate state of $(N - N_0)$ particles, and vice versa, due to particle number conservation – in agreement with Eq. (5.2). Each non-condensate state $\hat{\rho}_\perp(N - N_0, T)$ of $(N - N_0)$ particles is a thermal mixture projected onto the subspace of $(N - N_0)$ particles,

$$\hat{\rho}_\perp(N - N_0, T) = \frac{\hat{\mathscr{Q}}_{N-N_0} e^{-\beta \hat{\mathcal{H}}_\perp} \hat{\mathscr{Q}}_{N-N_0}}{\mathscr{Z}_\perp(N - N_0)} , \qquad (5.12)$$

where $\hat{\mathscr{Q}}_{N-N_0}$ denotes the projector onto the non-condensate subspace $\mathscr{F}_\perp(N - N_0)$ (see Section 4.4). The normalization factor $\mathscr{Z}_\perp(N - N_0)$ corresponds to the partition function [10] of $(N - N_0)$ indistinguishable non-condensate particles, given by

$$\mathscr{Z}_\perp(N - N_0) = \mathrm{Tr}_{\mathscr{F}_\perp}\left\{\hat{\mathscr{Q}}_{N-N_0} e^{-\beta \hat{\mathcal{H}}_\perp} \hat{\mathscr{Q}}_{N-N_0}\right\} . \qquad (5.13)$$

The non-condensate Hamiltonian $\hat{\mathcal{H}}_\perp$ is diagonalized according to Section 4.3, by expanding the non-condensate field in the basis $\{|\Psi_k\rangle, k \in \mathbb{N}\}$. Hence the spectral decomposition of $\hat{\mathcal{H}}_\perp$ leads to the statistical mixture:

$$\hat{\rho}_\perp(N-N_0, T) = \hat{\mathcal{D}}_{N-N_0} \frac{\bigotimes_{k\neq 0} \sum_{N_k=0}^{\infty} p_k(N_k, T)|N_k\rangle\langle N_k|}{\mathcal{Z}_\perp(N-N_0)} \hat{\mathcal{D}}_{N-N_0} . \tag{5.14}$$

In Eq. (5.14), $N_k$ are occupation numbers of non-condensate single particle states $|\Psi_k\rangle$ and $p_k(N_k, T) = \exp[-\beta\epsilon_k N_k]$ are Boltzmann probability factors for indistinguishable particles. The corresponding partition function $\mathcal{Z}_\perp(N-N_0)$ of $(N-N_0)$ particles turns into:

$$\mathcal{Z}_\perp(N-N_0) = \sum_{\{N_k\}}^{(N-N_0)} \exp\left[-\beta \sum_{k\neq 0} \epsilon_k N_k\right] . \tag{5.15}$$

In Eq. (5.15), we have introduced the partial sum $\sum_{\{N_k\}}^{(N-N_0)}$, which denotes a summation over all partitions $N_1 = 0\ldots\infty, N_2 = 0\ldots\infty, \ldots : \sum_{k\neq 0} N_k = (N-N_0)$.

### 5.2.2 Evolution equation for the condensate

The partial trace of $\hat{\sigma}^{(N)}(t)$ over the non-condensate subspace $\mathcal{F}_\perp$ defines the reduced condensate density matrix $\hat{\rho}_0^{(N)}(t)$ of the state of $N$ particles,

$$\hat{\rho}_0^{(N)}(t) = \text{Tr}_{\mathcal{F}_\perp}\{\hat{\sigma}^{(N)}(t)\} = \sum_{N_0=0}^{N} p_N(N_0, t)|N_0\rangle\langle N_0| , \tag{5.16}$$

with $\text{Tr}_{\mathcal{F}_0}\hat{\rho}_0(t) = 1$. As already mentioned, also the total $N$-body state is completely described in terms of the condensate particle number distribution $p_N(N_0, t)$, see Eq. (5.11), under the assumption of rapid thermalization in the non-condensate. This allows us to derive a closed evolution equation for the reduced condensate density matrix, $\hat{\rho}_0(t)$, within the Markov dynamics assumption by inserting the thermalized state, Eq. (5.11), into the $N$-body evolution Eq. (5.9). Finally, taking the partial trace over the non-condensate leads to:

$$\Delta\hat{\rho}_0^{(N,I)}(t) = -\frac{i}{\hbar}\sum_{N_0=0}^{N}\int_t^{t+\Delta t} dt'\, \mathrm{Tr}_{\mathscr{F}_\perp}\left[\hat{V}_{0\perp}(t'), p_N(N_0,t)|N_0\rangle\langle N_0| \otimes \hat{\rho}_\perp(N-N_0,T)\right]$$
$$-\sum_{N_0=0}^{N}\int_t^{t+\Delta t} dt' \int_t^{t'} \frac{dt''}{\hbar^2}\, \mathrm{Tr}_{\mathscr{F}_\perp}\left[\hat{V}_{0\perp}(t'),\left[\hat{V}_{0\perp}(t''), p_N(N_0,t)|N_0\rangle\langle N_0| \otimes \hat{\rho}_\perp(N-N_0,T)\right]\right]. \quad (5.17)$$

At this stage, we can verify that, according to Eq. (5.17), the time evolution of the N-body state $\hat{\sigma}^{(N)}(\Delta t)$ is completely determined by $\hat{\rho}_0^{(N)}(\Delta t)$, such that the thermalization ansatz in Eq. (5.11) is valid for any iterative step $\Delta t$, and therefore in fact for all times during the condensate formation process, reflecting the Born-Markov dynamics assumption for a dilute Bose-Einstein condensate.

## 5.3 Contribution of first order interaction terms

In Eq. (5.17), the contribution of the individual interaction terms for single particle ($\leadsto$), pair ($\leadsto\!\!\!\leadsto$) and scattering ($\circlearrowright$) processes are taken into account by the interaction term $\hat{V} = \hat{V}_{\leadsto} + \hat{V}_{\leadsto\!\!\!\leadsto} + \hat{V}_{\circlearrowright}$. First, we will verify that all contributions linear in the interaction operator $\hat{V}_{0\perp}$ vanish, before the evolution equations for the second order terms in $g$ are derived from Eq. (5.17).

### 5.3.1 General operator averages in the Bose state

First, we introduce a useful identity for operator averages, frequently applied in the derivation of the condensate quantum master equation. For the N-body state $\hat{\sigma}^{(N)}(t)$ in Eq. (5.11), the partial trace of an operator of the form $\mathscr{S} \otimes \hat{\mathscr{E}} : \mathscr{F}_0 \otimes \mathscr{F}_\perp \to \mathscr{F}_0 \otimes \mathscr{F}_\perp$, where $\mathscr{S} : \mathscr{F}_0 \to \mathscr{F}_0$ acts solely on the condensate Fock space, and $\hat{\mathscr{E}} : \mathscr{F}_\perp \to \mathscr{F}_\perp$ solely on the non-condensate Fock space, adopts the form

$$\mathrm{Tr}_{\mathscr{F}_\perp}\left\{\left(\mathscr{S} \otimes \hat{\mathscr{E}}\right)\hat{\sigma}^{(N,I)}(t)\right\} = \sum_{N_0=0}^{N} p_N(N_0,t)\left(\mathscr{S}|N_0\rangle\langle N_0|\right)\mathrm{Tr}_{\mathscr{F}_\perp}\left\{\hat{\mathscr{E}}\hat{\rho}_\perp(N-N_0,T)\right\}. \quad (5.18)$$

| $j$ | $\mathscr{P}_j(\vec{r},t)$ | $\mathscr{P}_j^\dagger(\vec{r},t)$ | $\hat{\mathscr{E}}_j(\vec{r},t)$ | $\hat{\mathscr{E}}_j^\dagger(\vec{r},t)$ |
|---|---|---|---|---|
| $\rightsquigarrow$ | $g\hat{\Psi}_0(\vec{r},t)$ | $g\hat{\Psi}_0^\dagger(\vec{r},t)$ | $\hat{\Psi}_\perp^\dagger(\vec{r},t)\hat{\Psi}_\perp(\vec{r},t)\hat{\Psi}_\perp(\vec{r},t)$ | $\hat{\Psi}_\perp^\dagger(\vec{r},t)\hat{\Psi}_\perp^\dagger(\vec{r},t)\hat{\Psi}_\perp(\vec{r},t)$ |
| $\leftrightsquigarrow$ | $g\hat{\Psi}_0(\vec{r},t)\hat{\Psi}_0(\vec{r},t)$ | $g\hat{\Psi}_0^\dagger(\vec{r},t)\hat{\Psi}_0^\dagger(\vec{r},t)$ | $\hat{\Psi}_\perp^\dagger(\vec{r},t)\hat{\Psi}_\perp^\dagger(\vec{r},t)$ | $\hat{\Psi}_\perp(\vec{r},t)\hat{\Psi}_\perp(\vec{r},t)$ |
| $\circlearrowright$ | $g\hat{\Psi}_0^\dagger(\vec{r},t)\hat{\Psi}_0(\vec{r},t)$ | $g\hat{\Psi}_0^\dagger(\vec{r},t)\hat{\Psi}_0(\vec{r},t)$ | $\hat{\Psi}_\perp^\dagger(\vec{r},t)\hat{\Psi}_\perp(\vec{r},t)$ | $\hat{\Psi}_\perp^\dagger(\vec{r},t)\hat{\Psi}_\perp(\vec{r},t)$ |

Table 5.1: *The formal structure of each interaction term $\hat{V}_j$ is identical for single particle ($j = \rightsquigarrow$), pair ($j = \leftrightsquigarrow$) and scattering ($j = \circlearrowright$) processes. The operators $\mathscr{P}_j(\vec{r},t)$, $\mathscr{P}_j^\dagger(\vec{r},t)$, $\hat{\mathscr{E}}_j(\vec{r},t)$ and $\hat{\mathscr{E}}_j^\dagger(\vec{r},t)$, have thus to be substituted in Eq. (5.21) for each process according to the table.*

Moreover, the abbreviation

$$\left\langle \hat{\mathscr{E}} \right\rangle_{\mathscr{F}_\perp}^{(N-N_0)} \equiv \mathrm{Tr}_{\mathscr{F}_\perp}\left\{ \hat{\mathscr{E}}\hat{\rho}_\perp(N-N_0,T) \right\} = \mathscr{Z}_\perp^{-1}(N-N_0)\mathrm{Tr}_{\mathscr{F}_\perp(N-N_0)}\left\{ \hat{\mathscr{E}}e^{-\beta\hat{H}_\perp} \right\}, \quad (5.19)$$

denotes the average of an operator $\hat{\mathscr{E}}: \mathscr{F}_\perp \to \mathscr{F}_\perp$ over the non-condensate subspace $\mathscr{F}_\perp$ with respect to the thermal non-condensate state $\hat{\rho}_\perp(N-N_0,T)$ of $(N-N_0)$ particles, see Eq. (5.14).

### 5.3.2 Vanishing of linear interaction terms

Linear and nonlinear contributions in $g$ in Eq. (5.17) need to be determined separately. With the decomposition of the interaction term $\hat{V}_{0\perp}^{(l)}(t') = \hat{V}_\rightsquigarrow^{(l)}(t') + \hat{V}_\leftrightsquigarrow^{(l)}(t') + \hat{V}_\circlearrowright^{(l)}(t')$, first order terms in Eq. (5.17) can be rewritten as

$$\mathrm{Tr}_{\mathscr{F}_\perp}\left[\hat{V}_{0\perp}^{(l)}(t'), \hat{\sigma}^{(N,l)}(t)\right] = \sum_j \mathrm{Tr}_{\mathscr{F}_\perp}\left[\hat{V}_j^{(l)}(t'), \hat{\sigma}^{(N,l)}(t)\right], \quad (5.20)$$

where $j = \rightsquigarrow, \leftrightsquigarrow, \circlearrowright$. Furthermore, each of the different interaction terms $\hat{V}_\rightsquigarrow^{(l)}(t), \hat{V}_\leftrightsquigarrow^{(l)}(t)$

## 5.3. CONTRIBUTION OF FIRST ORDER INTERACTION TERMS

and $\hat{\mathcal{V}}_\circlearrowleft^{(I)}(t)$ can be formally decomposed as

$$\hat{\mathcal{V}}_j^{(I)}(t') = \int_{\mathscr{C}} d\vec{r} \left[ \hat{\mathscr{S}}_j^\dagger(\vec{r},t') \otimes \hat{\mathscr{E}}_j(\vec{r},t') + \hat{\mathscr{S}}_j(\vec{r},t') \otimes \hat{\mathscr{E}}_j^\dagger(\vec{r},t') \right], \quad (5.21)$$

where the operators $\hat{\mathscr{S}}_j(\vec{r},t') : \mathscr{F}_0 \to \mathscr{F}_0$ and $\hat{\mathscr{E}}_j(\vec{r},t') : \mathscr{F}_\perp \to \mathscr{F}_\perp$ represent the different field operators in the interaction picture, see Table 5.1, with $j = \rightsquigarrow, \leftrightsquigarrow$ and $\circlearrowleft$, acting on the condensate and non-condensate subspaces $\mathscr{F}_0$ and $\mathscr{F}_\perp$ separately.

For any $\hat{\mathcal{V}}_j^{(I)}(t)$ in Eq. (5.21) the operator average in Eq. (5.18) yields

$$\begin{aligned}\mathrm{Tr}_{\mathscr{F}_\perp}\left[\hat{\mathcal{V}}_j(t'),\hat{\sigma}^{(N,I)}(t)\right] = &\int_{\mathscr{C}} d\vec{r} \sum_{N_0=0}^N \left[\hat{\mathscr{S}}_j^\dagger(\vec{r},t'), p_N(N_0,t)|N_0\rangle\langle N_0|\right]\left\langle\hat{\mathscr{E}}_j(\vec{r},t')\right\rangle_{\mathscr{F}_\perp}^{(N-N_0)}\\ &+ \int_{\mathscr{C}} d\vec{r} \sum_{N_0=0}^N \left[\hat{\mathscr{S}}_j(\vec{r},t'), p_N(N_0,t)|N_0\rangle\langle N_0|\right]\left\langle\hat{\mathscr{E}}_j^\dagger(\vec{r},t')\right\rangle_{\mathscr{F}_\perp}^{(N-N_0)}.\end{aligned} \quad (5.22)$$

For single particle interaction processes described by $\hat{\mathcal{V}}_\rightsquigarrow^{(I)}(t')$, i.e., $j=\rightsquigarrow$ in Eq. (5.21), first order contributions vanish as a result of particle number conservation,

$$\mathrm{Tr}_{\mathscr{F}_\perp}\left[\hat{\mathcal{V}}_\rightsquigarrow^{(I)}(t'),\hat{\sigma}^{(N,I)}(t)\right] = 0, \quad (5.23)$$

which can be directly checked by setting the corresponding single particle interaction terms of Table 5.1 for the operators $\hat{\mathscr{S}}_\rightsquigarrow(\vec{r},t')$ and $\hat{\mathscr{E}}_\rightsquigarrow(\vec{r},t')$, respectively, as well as their hermitian conjugates, into Eq. (5.22). This is a direct consequence of the fact that $\hat{\mathcal{V}}_\rightsquigarrow$ changes the number of non-condensate particles from initially $(N-N_0)$ to $(N-N_0\pm 1)$, and hence:

$$\left\langle\hat{\Psi}_\perp^\dagger(\vec{r},t')\hat{\Psi}_\perp(\vec{r},t')\hat{\Psi}_\perp(\vec{r},t')\right\rangle_{\mathscr{F}_\perp}^{(N-N_0)} = \left\langle\hat{\Psi}_\perp^\dagger(\vec{r},t')\hat{\Psi}_\perp^\dagger(\vec{r},t')\hat{\Psi}_\perp(\vec{r},t')\right\rangle_{\mathscr{F}_\perp}^{(N-N_0)} = 0. \quad (5.24)$$

The same argument applies also to the first order contributions for pair events described by $\hat{\mathcal{V}}^{(l)}_{\rightsquigarrow\rightsquigarrow}(t')$, i.e., $j = \rightsquigarrow\rightsquigarrow$ in Eq. (5.21). They too do not contribute to the master equation,

$$\mathrm{Tr}_{\mathcal{F}_\perp}\left[\hat{\mathcal{V}}^{(l)}_{\rightsquigarrow\rightsquigarrow}(t'), \hat{\sigma}^{(N,l)}(t)\right] = 0 , \qquad (5.25)$$

because again, due to the changing number of non-condensate particles,

$$\left\langle \hat{\Psi}^\dagger_\perp(\vec{r},t')\hat{\Psi}^\dagger_\perp(\vec{r},t')\right\rangle^{(N-N_0)}_{\mathcal{F}_\perp} = \left\langle \hat{\Psi}_\perp(\vec{r},t')\hat{\Psi}_\perp(\vec{r},t')\right\rangle^{(N-N_0)}_{\mathcal{F}_\perp} = 0 . \qquad (5.26)$$

In contrast, scattering terms $\hat{\mathcal{V}}^{(l)}_\circlearrowleft(t')$ lead to non-vanishing non-condensate expectation values from first order terms, $\left\langle \hat{\Psi}^\dagger_\perp(\vec{r},t')\hat{\Psi}_\perp(\vec{r},t')\right\rangle^{(N-N_0)}_{\mathcal{F}_\perp} \neq 0$. Nevertheless, they neither arise in the master equation, because the term $\hat{\mathcal{V}}_\circlearrowleft$ leaves the condensate and non-condensate occupation in the diagonal state $\hat{\sigma}^{(N)}(t)$ invariant, which entails:

$$\mathrm{Tr}_{\mathcal{F}_\perp}\left[\hat{\mathcal{V}}^{(l)}_\circlearrowleft(t'), \hat{\sigma}^{(N,l)}(t)\right] = 0 . \qquad (5.27)$$

Since all first order terms vanish, it remains to evaluate second order terms in the interaction term $\hat{\mathcal{V}}_{01}$ in Eq. (5.17) for single particle ($\rightsquigarrow$), pair ($\rightsquigarrow\rightsquigarrow$) and scattering ($\circlearrowleft$) processes.

## 5.4 Dynamical separation of two body interaction terms

As shown in Section 5.3.2, first order contributions vanish due to the rapid equilibration in the non-condensate and particle number conservation. Thus, the evolution Eq. (5.17) reduces to a sum over second order contributions in $\hat{\mathcal{V}}_{01} = \hat{\mathcal{V}}_{\rightsquigarrow} + \hat{\mathcal{V}}_{\rightsquigarrow\rightsquigarrow} + \hat{\mathcal{V}}_\circlearrowleft$:

$$\Delta\hat{\rho}^{(N,l)}_0(t) = -\sum_{N_0=0}^{N}\sum_{i,j}\int_t^{t+\Delta t}dt'\int_t^{t'}\frac{dt''}{\hbar^2}\mathrm{Tr}_{\mathcal{F}_\perp}\left[\hat{\mathcal{V}}^{(l)}_i(t'),\left[\hat{\mathcal{V}}^{(l)}_j(t''), p_N(N_0,t)|N_0\rangle\langle N_0|\otimes\hat{\rho}_\perp(N-N_0,T)\right]\right] , \qquad (5.28)$$

with $i,j = \rightsquigarrow, \looparrowright, \circlearrowleft$. Moreover, any mixed commutator in Eq. (5.28) is zero, meaning that single particle, pair and scattering events in the gas occur independently. This is shown by either brute force calculation, or by recognizing that there exists no combination of *two mixed* ($i \neq j$) two body diagrams (and hermitian conjugates of them) depicted in Fig. 5.1, which conserves the total particle number $N$. This is due to the fact that the three types of interaction processes ($\rightsquigarrow, \looparrowright, \circlearrowleft$) refer to different particle number changes $\Delta N_0 = -\Delta N_\perp = 0, \pm 1, \pm 2$, respectively.

Since the density matrix exhibits no coherences between states with different particle numbers, the trace over these particle number breaking terms is zero. Only conjugate two body interaction diagrams ($i = j$) of the same type in Fig. 5.1 therefore contribute to the master equation. Single particle, pair and scattering processes consequently occur as dynamically independent:

$$\frac{\Delta \hat{\rho}_0^{(N,l)}(t)}{\Delta t} = \left.\frac{\Delta \hat{\rho}_0^{(N,l)}(t)}{\Delta t}\right|_{\rightsquigarrow} + \left.\frac{\Delta \hat{\rho}_0^{(N,l)}(t)}{\Delta t}\right|_{\looparrowright} + \left.\frac{\Delta \hat{\rho}_0^{(N,l)}(t)}{\Delta t}\right|_{\circlearrowleft}, \qquad (5.29)$$

where we introduced the abbreviation

$$\left.\frac{\Delta \hat{\rho}_0^{(N,l)}(t)}{\Delta t}\right|_j \equiv -\sum_{N_0=0}^{N} \int_t^{t+\Delta t} dt' \int_t^{t'} \frac{dt''}{\hbar^2 \Delta t} \operatorname{Tr}_{\mathscr{F}_\perp} \left[ \hat{\mathcal{V}}_j(t'), \left[ \hat{\mathcal{V}}_j(t''), p_N(N_0, t) | N_0 \rangle \langle N_0 | \otimes \hat{\rho}_\perp(N - N_0, T) \right] \right], \qquad (5.30)$$

with $j = \rightsquigarrow, \looparrowright$ and $\circlearrowleft$.

## 5.5 Lindblad operators and transition rates

Using the decomposition of the different interaction terms $\hat{\mathcal{V}}_j^{(l)}(t)$ in Eq. (5.21) for $j = \rightsquigarrow, \looparrowright$ and $\circlearrowleft$, with corresponding field operators $\hat{\mathscr{S}}_j(\vec{r}, t'), \hat{\mathscr{S}}_j^\dagger(\vec{r}, t') : \mathscr{F}_0 \to \mathscr{F}_0$ and $\hat{\mathscr{E}}_j(\vec{r}, t'), \hat{\mathscr{E}}_j^\dagger(\vec{r}, t') : \mathscr{F}_\perp \to \mathscr{F}_\perp$ according to Table 5.1, we further derive the master equation for each coarse-grained evolution term in Eq. (5.30), $j = \rightsquigarrow, \looparrowright, \circlearrowleft$, in a rather

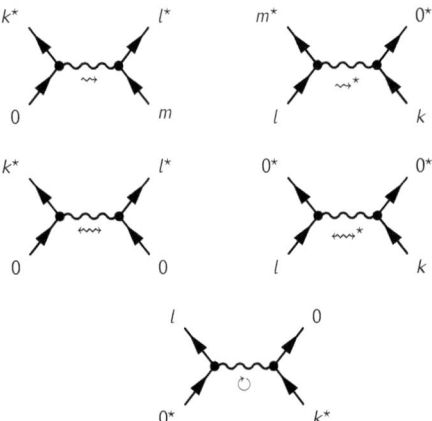

Figure 5.1: *Correlations between different types of diagrams for single particle loss (⇝) and feeding (⇝*) processes, pair loss (⟿) and feeding (⟿*) processes and scattering processes (○) cancel out, since any combination of mixed diagrams violates particle number conservation, as formally expressed by Eq. (5.30). Evolution terms of the N-body state relating to single particle, pair and scattering processes are therefore independent. In contrast, all particle interactions following the same diagrams will coherently interfere below $T_c$, leading to enhanced particle feedings (and losses) of the output channels.*

lengthy calculation which is skipped here. Herefore, we use that the averages over $\hat{\mathscr{E}}_j^\dagger(\vec{r},t')\hat{\mathscr{E}}_j^\dagger(\vec{r}\,',t'')$ and $\hat{\mathscr{E}}_j(\vec{r},t')\hat{\mathscr{E}}_j(\vec{r}\,',t'')$ are zero for all different interaction terms $j = \leadsto, \rightsquigarrow, \circlearrowright$, because they do not conserve the number of particles:

$$\left\langle \hat{\mathscr{E}}_j^\dagger(\vec{r},t')\hat{\mathscr{E}}_j^\dagger(\vec{r}\,',t'') \right\rangle_{\mathscr{F}_\perp}^{(N-N_0)} = \left\langle \hat{\mathscr{E}}_j(\vec{r},t')\hat{\mathscr{E}}_j(\vec{r}\,',t'') \right\rangle_{\mathscr{F}_\perp}^{(N-N_0)} = 0 \,, \qquad (5.31)$$

for $j = \leadsto, \rightsquigarrow, \circlearrowright$. Thus, only the remaining averages of the non-condensate field need to be treated, i.e. averages over operator products of the form $\hat{\mathscr{E}}_j(\vec{r},t')\hat{\mathscr{E}}_j^\dagger(\vec{r}\,',t'')$ and $\hat{\mathscr{E}}_j^\dagger(\vec{r},t')\hat{\mathscr{E}}_j(\vec{r}\,',t'')$. Thereby, the following two time averages are obtained, called two point correlation functions for the non-condensate field in normal order,

## 5.5. LINDBLAD OPERATORS AND TRANSITION RATES

$$\mathcal{G}_j^{(+)}(\vec{r},\vec{r}\,',N-N_0,T,\tau) = \left\langle \hat{\mathscr{E}}_j^\dagger(\vec{r},\tau)\hat{\mathscr{E}}_j(\vec{r}\,',0) \right\rangle_{\mathscr{F}_\perp}^{(N-N_0)}, \tag{5.32}$$

and, respectively, correlation functions of the non-condensate field in anti-normal order,

$$\mathcal{G}_j^{(-)}(\vec{r},\vec{r}\,',N-N_0,T,\tau) = \left\langle \hat{\mathscr{E}}_j(\vec{r},\tau)\hat{\mathscr{E}}_j^\dagger(\vec{r}\,',0) \right\rangle_{\mathscr{F}_\perp}^{(N-N_0)}, \tag{5.33}$$

where $\tau = t' - t''$ is the time difference between $t'$ and $t''$. The correlation functions depend on $\vec{r}, \vec{r}\,'$ and $\tau$, representing the coherent part of the non-condensate time evolution with respect to the linearized Hamiltonian $\hat{\mathcal{H}}_\perp$. These coherent parts obey the condition

$$\left(\mathcal{G}_j^{(\pm)}(\vec{r},\vec{r}\,',N-N_0,T,\tau)\right)^* = \mathcal{G}_j^{(\mp)}(\vec{r},\vec{r}\,',N-N_0,T,\tau) = \mathcal{G}_j^{(\pm)}(\vec{r},\vec{r}\,',N-N_0,T,-\tau). \tag{5.34}$$

Changing [20, 77] variables of integration, switching from $t'$ and $t''$ to $\tau$ and $t'$, leads to

$$\int_t^{t+\Delta t} dt' \int_t^{t'} dt'' = \int_0^{\Delta t} d\tau \int_{t+\tau}^{t+\Delta t} dt'. \tag{5.35}$$

According to the fact that the non-condensate thermalizes on the time scale $\tau_{\text{col}}$ due to atomic interactions [59, 63, 64], the correlation functions of the non-condensate field decay on a time scale $\tau_{\text{col}}$ [59].

To implement this irreversible decay, we add the exponential function $\exp[-\Gamma^2\tau^2]$ with $\Gamma = \tau_{\text{col}}^{-1}$. As a consequence of $\Delta t > \tau_{\text{col}}$, the time domain of integration over $d\tau$ can be extended from $\Delta t$ to $\infty$. The same is done for the domain of the integral over $t'$ by setting the lower bound $t + \tau$ to $t$, Thereby, each evolution term (for

## Chapter 5. LINDBLAD MASTER EQUATION FOR A BOSE-EINSTEIN CONDENSATE

$j = \leadsto, \mathrel{\leadsto\mkern-14mu\leadsto}, \circlearrowleft)$ in Eq. (5.30) turns into:

$$\left.\frac{\Delta \hat{\rho}_0^{(N,l)}(t)}{\Delta t}\right|_j = \sum_{N_0=0}^{N} \int_0^\infty d\tau \frac{e^{-\Gamma^2\tau^2}}{\hbar^2} \int_{\mathscr{C}\times\mathscr{C}} d\vec{r}\, d\vec{r}\,'\, \mathscr{S}_j(\vec{r},t) \mathscr{S}_j^\dagger(\vec{r}\,',t-\tau) \mathscr{X}_{N_0}^t \mathscr{G}_j^{(+)}(\vec{r},\vec{r}\,',N-N_0,T,\tau)$$

$$- \sum_{N_0=0}^{N} \int_0^\infty d\tau \frac{e^{-\Gamma^2\tau^2}}{\hbar^2} \int_{\mathscr{C}\times\mathscr{C}} d\vec{r}\, d\vec{r}\,'\, \mathscr{S}_j^\dagger(\vec{r},t) \mathscr{X}_{N_0}^t \mathscr{S}_j(\vec{r}\,',t-\tau) \mathscr{G}_j^{(+)}(\vec{r},\vec{r}\,',N-N_0,T,\tau)$$

$$+ \sum_{N_0=0}^{N} \int_0^\infty d\tau \frac{e^{-\Gamma^2\tau^2}}{\hbar^2} \int_{\mathscr{C}\times\mathscr{C}} d\vec{r}\, d\vec{r}\,'\, \mathscr{X}_{N_0}^t \mathscr{S}_j^\dagger(\vec{r}\,',t-\tau) \mathscr{S}_j(\vec{r},t) \mathscr{G}_j^{(-)}(\vec{r},\vec{r}\,',N-N_0,T,\tau)$$

$$- \sum_{N_0=0}^{N} \int_0^\infty d\tau \frac{e^{-\Gamma^2\tau^2}}{\hbar^2} \int_{\mathscr{C}\times\mathscr{C}} d\vec{r}\, d\vec{r}\,'\, \mathscr{S}_j(\vec{r}\,',t-\tau) \mathscr{X}_{N_0}^t \mathscr{S}_j^\dagger(\vec{r},t) \mathscr{G}_j^{(-)}(\vec{r},\vec{r}\,',N-N_0,T,\tau)$$

$$+ \text{h.c.} \,,$$

(5.36)

where we introduced $\mathscr{X}_{N_0}^t \equiv p_N(N_0,t)|N_0\rangle\langle N_0|$ for brevity. The occuring Lindblad type structure [85] is typical for Markov processes [19, 21], and can be already identified at this level of the master equation. Different Lindblad superoperators being related to single particle ($\leadsto$), pair ($\mathrel{\leadsto\mkern-14mu\leadsto}$) and scattering ($\circlearrowleft$) events are derived in the following for each process in the next subsections.

### 5.5.1 Lindblad evolution term for single particle processes ($\leadsto$)

In this section, the evolution Eq. (5.36) is worked out for $j = \leadsto$, yielding the transition rates associated to the two point correlation functions $\mathscr{G}_{\leadsto}^{(\pm)}(\vec{r},\vec{r}\,',N-N_0,T,\tau)$. Single particle events have been specified in Section 4.2.3 as two body interaction terms which annihilate, or create, one particle in the condensate mode $|\Psi_0\rangle$. The interaction term $\hat{V}_{\leadsto}^{(l)}(\vec{r},t)$ hence contains the sum of all diagrams

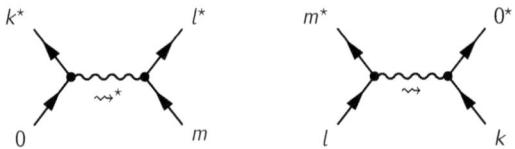

## 5.5. LINDBLAD OPERATORS AND TRANSITION RATES

see Fig. 5.1, refering to the field operators $\mathscr{S}_{\leadsto}(\vec{r},t)$ and $\hat{\mathscr{E}}_{\leadsto}(\vec{r},t)$, as well as to their hermitian conjugates, $\mathscr{S}_{\leadsto}^{\dagger}(\vec{r},t)$ and $\hat{\mathscr{E}}_{\leadsto}^{\dagger}(\vec{r},t)$ in Eq. (5.36) according to Table 5.1.

The time evolution of the coarse-grained evolution term in Eq. (5.36) due to single particle events, $j = \leadsto$, is specified by the normally ordered two point correlation function of the non-condensate field, $\mathscr{G}_{\leadsto}^{(+)}(\vec{r},\vec{r}\,',N-N_0,T,\tau)$ in Eq. (5.32),

$$\mathscr{G}_{\leadsto}^{(+)}(\vec{r},\vec{r}\,',N-N_0,T,\tau) = \left\langle \hat{\Psi}_{\perp}^{\dagger}(\vec{r},\tau)\hat{\Psi}_{\perp}^{\dagger}(\vec{r},\tau)\hat{\Psi}_{\perp}(\vec{r},\tau)\hat{\Psi}_{\perp}^{\dagger}(\vec{r}\,',0)\hat{\Psi}_{\perp}(\vec{r}\,',0)\hat{\Psi}_{\perp}(\vec{r}\,',0) \right\rangle_{\mathscr{F}_{\perp}}^{(N-N_0)}, \quad (5.37)$$

and by the anti-normally ordered counterpart,

$$\mathscr{G}_{\leadsto}^{(-)}(\vec{r},\vec{r}\,',N-N_0,T,\tau) = \left\langle \hat{\Psi}_{\perp}^{\dagger}(\vec{r},\tau)\hat{\Psi}_{\perp}(\vec{r},\tau)\hat{\Psi}_{\perp}(\vec{r},\tau)\hat{\Psi}_{\perp}^{\dagger}(\vec{r}\,',0)\hat{\Psi}_{\perp}^{\dagger}(\vec{r}\,',0)\hat{\Psi}_{\perp}(\vec{r}\,',0) \right\rangle_{\mathscr{F}_{\perp}}^{(N-N_0)}, \quad (5.38)$$

correspondingly.[3]

The action of condensate field operators onto an element of the condensate Fock space, $\hat{\Psi}_0(\vec{r})|N_0\rangle = \sqrt{N_0}\Psi_0(\vec{r})|N_0-1\rangle$, and $\hat{\Psi}_0^{\dagger}(\vec{r})|N_0\rangle = \sqrt{N_0+1}\Psi_0^{*}(\vec{r})|N_0+1\rangle$, respectively, leads to the Lindblad term for microscopic single particle loss and feeding events, directly from Eq. (5.36). After multiplying $\hat{\mathcal{U}}_0^{\dagger}(t)$ from the left and $\hat{\mathcal{U}}_0(t)$ from the right, the latter equation turns into:

$$\hat{\mathcal{U}}_0^{\dagger}(t) \left.\frac{\Delta\hat{\rho}_0^{(N,l)}(t)}{\Delta t}\right|_{\leadsto} \hat{\mathcal{U}}_0(t) = \left\{ \sum_{N_0=0}^{N} p_N(N_0,t) \left( \hat{a}_0^{\dagger}|N_0\rangle\langle N_0|\hat{a}_0 - |N_0\rangle\langle N_0|\hat{a}_0\hat{a}_0^{\dagger} \right) \right\} \Lambda_{\leadsto}^{+}(N-N_0,T)$$

$$+ \left\{ \sum_{N_0=0}^{N} p_N(N_0,t) \left( \hat{a}_0|N_0\rangle\langle N_0|\hat{a}_0^{\dagger} - \hat{a}_0^{\dagger}\hat{a}_0|N_0\rangle\langle N_0| \right) \right\} \Lambda_{\leadsto}^{-}(N-N_0,T)$$

$$+ \text{h.c.} \ .$$

(5.39)

---

[3]The correlation functions $\mathscr{G}_{\leadsto}^{(\pm)}(\vec{r},\vec{r}\,',N-N_0,T,\tau)$ are calculated explicitly in Chapter 7.1.

The complex valued, time averaged rates $\Lambda^\pm_\leftrightarrow(N-N_0,T) \in \mathbb{C}$ are given by

$$\Lambda^\pm_\leftrightarrow(N-N_0,T) = \frac{g^2}{\hbar^2} \iint_{\mathscr{C}\times\mathscr{C}} d\vec{r}\,d\vec{r}^{\,\prime\prime}\,\Psi_0^*(\vec{r})\Psi_0(\vec{r}^{\,\prime\prime}) \int_0^\infty d\tau\, e^{\pm i\omega_0\tau - \Gamma^2\tau^2}\mathscr{G}^{(\pm)}_\leftrightarrow(\vec{r},\vec{r}^{\,\prime\prime},N-N_0,T,\tau) \,. \quad (5.40)$$

They depend (i) on the number of non-condensate particles $N-N_0$ via the two point correlation functions $\mathscr{G}^{(\pm)}_\leftrightarrow(\vec{r},\vec{r}^{\,\prime\prime},N-N_0,T,\tau)$ of non-condensate fields in Eqs. (5.37, 5.38), (ii) on the condensate wave function and its conjugate, $\Psi_0(\vec{r})$ and $\Psi_0^*(\vec{r})$, and (iii) on $\omega_0 = \mu_0/\hbar$, thus on the eigenvalue of the Gross-Pitaevskii equation (4.4). Furthermore, $T$ labels the final gas temperature and $g$ is the two particle interaction strength.

The rates $\Lambda^\pm_\leftrightarrow(N-N_0,T)$ in Eq. (5.40) are in general complex numbers: To separate the real from the complex part of the evolution Eq. (5.39) – the first appears as the real-time evolution, hence defining real valued single particle exchange rates between condensate and non-condensate, whereas the latter is associated to a shift of the condensate single particle energy (see Section 7.4) – the complex valued rate $\Lambda^\pm_\leftrightarrow(N-N_0,T)$ is decomposed into

$$\Lambda^\pm_\leftrightarrow(N-N_0,T) \equiv \lambda^\pm_\leftrightarrow(N-N_0,T) + i\,\Delta^\pm_\leftrightarrow(N-N_0,T) \,. \quad (5.41)$$

The real parts $\lambda^\pm_\leftrightarrow(N-N_0,T) = \mathscr{R}\{\Lambda^\pm_\leftrightarrow(N-N_0,T)\}$ in Eq. (5.40) are called single particle feeding and loss rates, given a state with $N_0$ particles populating the condensate mode in the Bose gas of $N$ particles. The imaginary part, $\Delta^\pm_\leftrightarrow(N-N_0,T) = \mathscr{I}\{\Lambda^\pm_\leftrightarrow(N-N_0,T)\}$, characterizes the single particle energy shift arising from virtual processes (see below).

The real part of evolution Eq. (5.39) for the reduced condensate density matrix $\hat{\rho}_0^{(N)}(t)$ originating from single particle processes is thus determined by the terms in Eq. (5.39) proportional to $\lambda^\pm_\leftrightarrow(N-N_0,T)$, after decomposition of the complex valued rates like in Eq. (5.41). Those contributions can now be described in terms of a

## 5.5. LINDBLAD OPERATORS AND TRANSITION RATES

superoperator $\mathscr{L}_\leadsto$ acting onto the subspace of the reduced condensate density matrix:

$$\mathscr{L}_\leadsto\left[\hat{\rho}_0^{(N)}(t)\right] = \sum_{N_0=0}^{N} \Gamma_N^+(N_0, T)\left[\hat{\mathscr{S}}_+(N_0)\hat{\rho}_0^{(N)}(t)\hat{\mathscr{S}}_+^\dagger(N_0) - \frac{1}{2}\left\{\hat{\mathscr{S}}_+^\dagger(N_0)\hat{\mathscr{S}}_+(N_0), \hat{\rho}_0^{(N)}(t)\right\}_+\right]$$
$$+ \sum_{N_0=0}^{N} \Gamma_N^-(N_0, T)\left[\hat{\mathscr{S}}_-(N_0)\hat{\rho}_0^{(N)}(t)\hat{\mathscr{S}}_-^\dagger(N_0) - \frac{1}{2}\left\{\hat{\mathscr{S}}_-^\dagger(N_0)\hat{\mathscr{S}}_-(N_0), \hat{\rho}_0^{(N)}(t)\right\}_+\right] ,$$
(5.42)

where $\{\hat{X}, \hat{Y}\}_+ \equiv \hat{X}\hat{Y} + \hat{Y}\hat{X}$ denotes the anti-commutator of the two operators $\hat{X}$ and $\hat{Y}$. Moreover, single particle condensate feeding and losses are described by rank one quantum jump operators $\hat{\mathscr{S}}_\pm(N_0)$ in Eq. (5.42), defined by

$$\hat{\mathscr{S}}_+(N_0) \equiv |N_0+1\rangle\langle N_0| ,$$
(5.43)

for quantum jumps of the condensate particle number $N_0 \to N_0+1$, and by

$$\hat{\mathscr{S}}_-(N_0) \equiv |N_0-1\rangle\langle N_0| ,$$
(5.44)

for quantum jumps $N_0 \to N_0-1$. They account for single particle condensate feeding (+) and loss (−) events, which are induced by the non-condensate environment. The jump operators satisfy the general relation for Kraus operators [19, 21],

$$\sum_{N_0=0}^{N} \hat{\mathscr{S}}_+^\dagger(N_0)\hat{\mathscr{S}}_+(N_0) = \sum_{N_0=0}^{N} \hat{\mathscr{S}}_-^\dagger(N_0)\hat{\mathscr{S}}_-(N_0) = \hat{1}_{\mathscr{F}_0} .$$
(5.45)

Equation (5.42) obeys the so called "Lindblad form" [85], which is characteristic for quantum Markov jump processes [19]. Remarkably, these jumps of particles into and out of the condensate, respectively, directly reflect the wave character of the particles in spatial representation. The effective condensate particle feedings and losses are quantified by the transition rates $\Gamma_N^+(N_0, T) = 2(N_0 + 1)\lambda_{\leftrightsquigarrow}^+(N - N_0, T)$ and $\Gamma_N^-(N_0, T) = 2N_0 \lambda_{\leftrightsquigarrow}^-(N - N_0, T)$ in Eqs. (5.40, 5.41, 5.42).

The imaginary part of the complex rate $\Lambda_{\leftrightsquigarrow}^\pm(N-N_0, T)$ in Eq. (5.40) generally leads to a coherent contribution to the Lindblad master equation, here given by

$$-i \sum_{N_0=0}^{N} \Delta_{\leftrightsquigarrow}(N - N_0, T) \left[ \hat{a}_0^\dagger \hat{a}_0, p_N(N_0, t)|N_0\rangle\langle N_0| \right] = 0 , \qquad (5.46)$$

with $\Delta_{\leftrightsquigarrow}(N_0, N - N_0, T) = \Delta_{\leftrightsquigarrow}^{(+)}(N_0, N - N_0, T) + \Delta_{\leftrightsquigarrow}^{(-)}(N_0, N - N_0, T)$, the net single particle energy shift[4]. The coherent evolution term in Eq. (5.46), however, vanishes and does not contribute to the time evolution of the reduced diagonal condensate density matrix. Consequently, the evolution term for single particle processes is fully captured by

$$\hat{u}_0^\dagger(t) \left. \frac{\Delta \hat{\rho}_0^{(N,l)}(t)}{\Delta t} \right|_{\leftrightsquigarrow} \hat{u}_0(t) = \mathscr{L}_{\leftrightsquigarrow} \left[ \hat{\rho}_0^{(N)}(t) \right] , \qquad (5.47)$$

with $\mathscr{L}_{\leftrightsquigarrow} \left[ \hat{\rho}_0^{(N)}(t) \right]$ defined by Eq. (5.42).

### 5.5.2 Lindblad evolution term for pair processes ($\leftrightsquigarrow$)

Pair events are specified as processes, where two condensate atoms are effectively created or annihilated. The diagrammatic representation of all different pair events is

---

[4] The energy shift $\Delta_{\leftrightsquigarrow}(N-N_0, T)$ can in principle be used to renormalize the condensate chemical potential $\mu_0$, in analogy to the Lamb shift in quantum optics [86, 87]. However, we verify in Section 7.4 that it is so small in dilute atomic gases (e.g., $\sim (0.0001 - 0.01)\hbar\omega$ in a three-dimensional isotropic harmonic trap) that we neglect the renormalization of $\mu_0$ for our present purpose.

## 5.5. LINDBLAD OPERATORS AND TRANSITION RATES

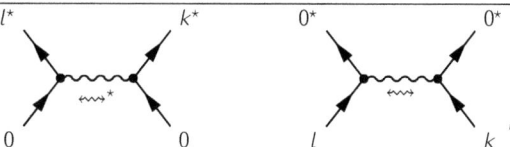

see Fig. 5.1.

Proceeding in compliance to the calculus for single particle processes, we insert the operators $\hat{\mathscr{S}}_{\rightsquigarrow}(\vec{r},t)$ and $\hat{\mathscr{E}}_{\rightsquigarrow}(\vec{r},t)$ as well as their hermitian conjugates according to Table 5.1 into Eq. (5.36), leading to a Lindblad evolution term for pair processes, $j=\rightsquigarrow$. The normally ordered two point correlation function $\mathscr{G}_{\rightsquigarrow}^{(+)}(\vec{r},\vec{r}\,',N-N_0,T,\tau)$ in Eq. (5.32) for pair events is given by

$$\mathscr{G}_{\rightsquigarrow}^{(+)}(\vec{r},\vec{r}\,',N-N_0,T,\tau) = \left\langle \hat{\Psi}_\perp^\dagger(\vec{r},\tau)\hat{\Psi}_\perp^\dagger(\vec{r},\tau)\hat{\Psi}_\perp(\vec{r}\,',0)\hat{\Psi}_\perp(\vec{r}\,',0) \right\rangle_{\mathscr{F}_\perp}^{(N-N_0)}, \quad (5.48)$$

whereas the anti-normally ordered pair correlation $\mathscr{G}_{\rightsquigarrow}^{(-)}(\vec{r},\vec{r}\,',N_0,T,\tau)$ function adopts the form

$$\mathscr{G}_{\rightsquigarrow}^{(-)}(\vec{r},\vec{r}\,',N_0,T,\tau) = \left\langle \hat{\Psi}_\perp(\vec{r},\tau)\hat{\Psi}_\perp(\vec{r},\tau)\hat{\Psi}_\perp^\dagger(\vec{r}\,',0)\hat{\Psi}_\perp^\dagger(\vec{r}\,',0) \right\rangle_{\mathscr{F}_\perp}^{(N-N_0)}, \quad (5.49)$$

and is obtained correspondingly by using Eq. (5.33).

The coarse-grained rate of variation for $j=\rightsquigarrow$ turns, after multiplication with $\hat{\mathcal{U}}_0^\dagger(t)$ from the left and with $\hat{\mathcal{U}}_0(t)$ from the right, into

$$\hat{\mathcal{U}}_0^\dagger(t)\left.\frac{\Delta\hat{\rho}_0^{(N,l)}}{\Delta t}\right|_{\rightsquigarrow}\hat{\mathcal{U}}_0(t) = \left\{ \sum_{N_0=0}^{N} p_N(N_0,t)\left(\hat{a}_0^\dagger|N_0\rangle\langle N_0|\hat{a}_0 - |N_0\rangle\langle N_0|\hat{a}_0\hat{a}_0^\dagger\right) \right\} \Lambda_{\rightsquigarrow}^+(N-N_0,T)$$

$$+ \left\{ \sum_{N_0=0}^{N} p_N(N_0,t)\left(\hat{a}_0|N_0\rangle\langle N_0|\hat{a}_0^\dagger - \hat{a}_0^\dagger\hat{a}_0|N_0\rangle\langle N_0|\right) \right\} \Lambda_{\rightsquigarrow}^-(N-N_0,T)$$

$$+ \text{h.c.}\,, \quad (5.50)$$

with pair annihilation and creation operators [88], $\hat{\alpha}_0 = \hat{a}_0 \hat{a}_0$ and $\hat{\alpha}_0^\dagger = \hat{a}_0^\dagger \hat{a}_0^\dagger$. The complex valued, time averaged pair feeding and loss rates are thus given by

$$\Lambda^\pm_{\sim\sim}(N-N_0, T) = \frac{g^2}{4\hbar^2} \iint_{\mathscr{C}\times\mathscr{C}} d\vec{r}\, d\vec{r}\,'' \Psi_0(\vec{r})\Psi_0(\vec{r})\Psi_0^*(\vec{r}\,'')\Psi_0^*(\vec{r}\,'') \int_0^\infty d\tau\, e^{\pm 2i\omega_0\tau - \Gamma^2\tau^2} \mathscr{G}^{(\pm)}_{\sim\sim}(\vec{r},\vec{r}\,'',N-N_0,T,\tau)\,,$$
(5.51)

with corresponding normally (and anti-normally) ordered pair two point correlation functions $\mathscr{G}^{(\pm)}_{\sim\sim}(\vec{r},\vec{r}\,'',N-N_0,T,\tau)$ of the non-condensate field in Eqs. (5.48, 5.49). They define the two body pair feeding and loss rate $\lambda^\pm_{\sim\sim}(N-N_0,T) = \mathscr{R}\{\Lambda^\pm_{\sim\sim}(N-N_0,T)\}$, and the two body pair energy shift $\Delta^{(\pm)}_{\sim\sim}(N-N_0,T) = \mathscr{I}\{\Lambda^\pm_{\sim\sim}(N-N_0,T)\}$, by means of the decomposition

$$\Lambda^\pm_{\sim\sim}(N-N_0,T) \equiv \lambda^\pm_{\sim\sim}(N-N_0,T) + i\,\Delta^\pm_{\sim\sim}(N-N_0,T)\,.$$
(5.52)

The Lindblad superoperator $\mathscr{L}_{\sim\sim}[\hat{\rho}_0^{(N)}(t)]$ which describes the pair dynamics governed by the real part of Eq. (5.50) is converted to the following form after the decomposition of pair feeding and loss rates in Eq. (5.52):

$$\mathscr{L}_{\sim\sim}\left[\hat{\rho}_0^{(N)}(t)\right] = \sum_{N_0=0}^{N} \gamma_N^-(N_0,T)\left[\mathscr{P}_+(N_0)\hat{\rho}_0^{(N)}(t)\mathscr{P}_+^\dagger(N_0) - \frac{1}{2}\left\{\mathscr{P}_+^\dagger(N_0)\mathscr{P}_+(N_0),\hat{\rho}_0^{(N)}(t)\right\}_+\right]$$
$$+ \sum_{N_0=0}^{N} \gamma_N^-(N_0,T)\left[\mathscr{P}_-(N_0)\hat{\rho}_0^{(N)}(t)\mathscr{P}_-^\dagger(N_0) - \frac{1}{2}\left\{\mathscr{P}_-^\dagger(N_0)\mathscr{P}_-(N_0),\hat{\rho}_0^{(N)}(t)\right\}_+\right]\,,$$
(5.53)

with pair quantum jump operators,

$$\mathscr{P}_+(N_0) \equiv |N_0+2\rangle\langle N_0|\,.$$
(5.54)

## 5.5. LINDBLAD OPERATORS AND TRANSITION RATES

for quantum jumps of the condensate particle number $N_0 \to N_0 + 2$, and with

$$\hat{\mathscr{P}}_-(N_0) \equiv |N_0 - 2\rangle\langle N_0| , \qquad (5.55)$$

for quantum jumps $N_0 \to N_0 - 2$, induced by the non-condensate environment. The form of Eq. (5.53) occurs also in two component chemical reactions [19], quantified here by the transition rates $\gamma_N^+(N_0, T) = 2\sqrt{(N_0+1)(N_0+2)}\lambda_{\sim\sim}^+(N-N_0, T)$ and $\gamma_N^-(N_0, T) = 2\sqrt{N_0(N_0-1)}\lambda_{\sim\sim}^-(N-N_0, T)$, respectively, defined via Eqs. (5.51, 5.52, 5.53). Also the pair jump operators $\hat{\mathscr{P}}_+(N_0)$ and $\hat{\mathscr{P}}_-(N_0)$ satisfy

$$\sum_{N_0=0}^{N} \hat{\mathscr{P}}_+^\dagger(N_0)\hat{\mathscr{P}}_+(N_0) = \sum_{N_0=0}^{N} \hat{\mathscr{P}}_-^\dagger(N_0)\hat{\mathscr{P}}_-(N_0) = \hat{1}_{\mathscr{F}_0} . \qquad (5.56)$$

Finally, the commutation relation $\left[\hat{a}_0, \hat{a}_0^\dagger\right] = 4(\hat{a}_0^\dagger \hat{a}_0 + \hat{1}/2)$ for pair operators leads to the coherent contribution of Eq. (5.50),

$$-i \sum_{N_0=0}^{N} \left[\Delta_{\sim\sim}(N-N_0, T)\hat{a}_0^\dagger \hat{a}_0 - 4\Delta_{\sim\sim}^+(N-N_0, T)\hat{a}_0^\dagger \hat{a}_0, p_N(N_0, t)|N_0\rangle\langle N_0|\right] = 0 . \qquad (5.57)$$

This contribution also vanishes exactly as a result of the diagonal reduced condensate density matrix, but nevertheless yields an estimate of energy shifts $\Delta_{\sim\sim}(N-N_0, T)$ induced by pair events. Again, the symbol $\Delta_{\sim\sim} = \Delta_{\sim\sim}^+ + \Delta_{\sim\sim}^-$ is used to label the net energy shift induced by pair events. Albeit this energy shift does not contribute to the master equation, it is interesting to estimate its order of magnitude, see Section 7.4.

The total evolution term for pair events,

$$\hat{\mathcal{U}}_0^\dagger(t) \left.\frac{\Delta \hat{\rho}_0^{(N,I)}(t)}{\Delta t}\right|_{\sim\sim} \hat{\mathcal{U}}_0(t) = \mathscr{L}_{\sim\sim}\left[\hat{\rho}_0^{(N)}(t)\right] , \qquad (5.58)$$

### 5.5.3 Evolution term for scattering processes (↻)

Finally, we verify that scattering events described by the interaction diagrams

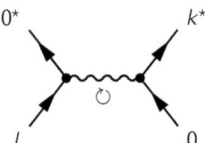

see Fig. 5.1, do not affect the number distribution $p_N(N_0, t)$ of condensate and non-condensate particles. Inserting the condensate and non-condensate field operators for $\hat{\mathscr{P}}_\circlearrowleft(\vec{r}, t)$ and $\hat{\mathscr{E}}_\circlearrowleft(\vec{r}, t)$ and their hermitian conjugates $\hat{\mathscr{P}}_\circlearrowleft^\dagger(\vec{r}, t)$ and $\hat{\mathscr{E}}_\circlearrowleft^\dagger(\vec{r}, t)$ into the evolution term in Eq. (5.36) according to Table 5.1 leads to only one two point correlation function of the non-condensate field,

$$\mathscr{G}_\circlearrowleft(\vec{r}, \vec{r}\,', N_0, T, \tau) = \left\langle \hat{\Psi}_\perp^\dagger(\vec{r}, \tau) \hat{\Psi}_\perp(\vec{r}, \tau) \hat{\Psi}_\perp^\dagger(\vec{r}\,', 0) \hat{\Psi}_\perp(\vec{r}\,', 0) \right\rangle_{\mathscr{F}_\perp}^{(N-N_0)}, \qquad (5.59)$$

see Eq. (5.32).

The evolution equation for the coarse-grained rate of variation for scattering processes hence turns into

$$\left. \frac{\Delta \hat{\rho}_0^{(N,l)}(t)}{\Delta t} \right|_\circlearrowleft = \left\{ \sum_{N_0=0}^{N} \hat{N}_0 p_N(N_0, t) |N_0\rangle\langle N_0| \hat{N}_0 - p_N(N_0, t) |N_0\rangle\langle N_0| \hat{N}_0^2 \right\} \Lambda_\circlearrowleft(N - N_0, T) + \text{h.c.} \equiv 0, \qquad (5.60)$$

where $\hat{N}_0 = \int_\mathscr{C} d\vec{r}\, \hat{\Psi}_0^\dagger(\vec{r}) \hat{\Psi}_0(\vec{r})$ is number operator of condensate particles. Even though Eq. (5.60) does not contribute to the master equation (leading to dephasing of the off-diagonal elements of the condensate density matrix), it yields the complex

## 5.6. QUANTUM MASTER EQUATION OF LINDBLAD TYPE

valued time averaged rate $\Lambda_\circlearrowleft(N-N_0,T)$ for atomic scattering processes between condensate and non-condensate particles with $\Delta N_0 = 0$:

$$\Lambda_\circlearrowleft(N-N_0,T) = \frac{4g^2}{\hbar^2} \iint_{\mathscr{C}\times\mathscr{C}} d\vec{r}^{\,*}\, d\vec{r}^{\,*\prime}\, |\Psi_0(\vec{r})|^2 |\Psi_0(\vec{r}^{\,\prime})|^2 \int_0^\infty d\tau\, e^{-\Gamma^2\tau^2} \mathscr{G}_\circlearrowleft(\vec{r}^{\,*},\vec{r}^{\,*\prime},N-N_0,T,\tau)\,, \quad (5.61)$$

with the corresponding correlation function $\mathscr{G}_\circlearrowleft(\vec{r}^{\,*},\vec{r}^{\,*\prime},N-N_0,T,\tau)$ for scattering events in Eq. (5.59). The decomposition of $\Lambda_\circlearrowleft(N-N_0,T) \in \mathbb{C}$ into

$$\Lambda_\circlearrowleft(N-N_0,T) \equiv \lambda_\circlearrowleft(N-N_0,T) + i\,\Delta_\circlearrowleft(N-N_0,T) \quad (5.62)$$

introduces the real valued scattering rate, $\lambda_\circlearrowleft(N-N_0,T) = \mathscr{R}\{\Lambda_\circlearrowleft(N-N_0,T)\}$, and the corresponding imaginary level shift, $\Delta_\circlearrowleft(N-N_0,T) = \mathscr{I}\{\Lambda_\circlearrowleft(N-N_0,T)\}$. Again, also the coherent contribution of scattering events is zero,

$$-i\sum_{N_0=0}^{N} \Delta_\circlearrowleft \left[\hat{a}_0^\dagger \hat{a}_0 \hat{a}_0^\dagger \hat{a}_0, p_N(N_0,t)|N_0\rangle\langle N_0|\right] = 0\,, \quad (5.63)$$

without any contribution due to scattering events:

$$\hat{\mathcal{U}}^\dagger(t)\left.\frac{\Delta\hat{\rho}_0^{(N,I)}(t)}{\Delta t}\right|_\circlearrowleft \hat{\mathcal{U}}(t) = 0\,, \quad (5.64)$$

The $N$-body state $\hat{\sigma}^{(N,I)}(t)$ in Eq. (5.11) therefore evolves in time only with respect to single particle ($\leadsto$) and pair ($\rightsquigarrow$) processes, described by the Lindblad terms in Eqs. (5.42, 5.53).

## 5.6 Quantum master equation of Lindblad type

The different results of the derivation in Sections 5.1 - 5.5 are now summarized in order to collect all relevant Lindblad terms for the dynamics of the reduced condensate state $\hat{\rho}_0^{(N)}(t)$. The relation between the reduced condensate density matrix in the Schrödinger and the interaction picture is

$$\frac{\Delta\hat{\rho}_0^{(N)}(t)}{\Delta t} = -\frac{i}{\hbar}\left[\hat{\mathcal{H}}_0,\hat{\rho}_0^{(N)}(t)\right] + \hat{\mathcal{U}}_0^\dagger(t)\frac{\Delta\hat{\rho}_0^{(N,I)}(t)}{\Delta t}\hat{\mathcal{U}}_0(t) \ . \tag{5.65}$$

Equation (5.65) formally contains the coherent evolution of $\hat{\rho}_0^{(N)}(t)$ with respect to the condensate Hamiltonian $\hat{\mathcal{H}}_0$, and the instantaneous rate of variation of the condensate density matrix in the interaction picture, $\partial_t\hat{\rho}_0^{(N,I)}(t)$. The Lindblad master equation simplifies further, because the Hamiltonian time evolution $[\hat{\mathcal{H}}_0,\hat{\rho}_0^{(N)}(t)]$ vanishes as a result of the condensate Hamiltonian $\hat{\mathcal{H}}_0$ being only proportional to the first and second order of the condensate number operator. It therefore commutes with the diagonal condensate state $\hat{\rho}_0^{(N)}(t)$.

The coarse-grained rate of variation [20], $\Delta\hat{\rho}_0^{(N,I)}(t)/\Delta t$, is the instantaneous rate of variation $\partial_t\hat{\rho}_0^{(N,I)}(t)$ averaged over the time interval $\Delta t$:

$$\left.\frac{\Delta\hat{\rho}_0^{(N,I)}(t)}{\Delta t}\right|_j = \left.\frac{\Delta\hat{\rho}_0^{(N,I)}(t+\Delta t) - \Delta\hat{\rho}_0^{(N,I)}(t)}{\Delta t}\right|_j = \frac{1}{\Delta t}\int_t^{t+\Delta t}dt'\left.\frac{\partial\hat{\rho}_0^{(N,I)}(t')}{\partial t'}\right|_j . \tag{5.66}$$

All rapid variations on time scales smaller than $\Delta t$ are washed out in this average. However, since the time interval $\Delta t$ is still much shorter than the condensate formation time $\tau_0$, the instantaneous rate of variation in Eq. (5.65) is well approximated by the coarse-grained evolution $\Delta\hat{\rho}_0^{(N)}(t)/\Delta t$ in Eq. (5.29), which entails

$$\frac{\partial\hat{\rho}_0^{(N,I)}(t)}{\partial t} \approx \frac{\Delta\hat{\rho}_0^{(N,I)}(t)}{\Delta t} = \left[\left.\frac{\Delta\hat{\rho}_0^{(N,I)}(t)}{\Delta t}\right|_{\leadsto} + \left.\frac{\Delta\hat{\rho}_0^{(N,I)}(t)}{\Delta t}\right|_{\rightsquigarrow} + \left.\frac{\Delta\hat{\rho}_0^{(N,I)}(t)}{\Delta t}\right|_{\circlearrowleft}\right] , \tag{5.67}$$

as explained in Section 5.4. Summarizing the explicit evolution terms for single particle processes ($\leadsto$) in Eq. (5.42), pair processes ($\rightsquigarrow$) in Eq. (5.53), and using the fact that scattering processes ($\circlearrowleft$) in Eq. (5.64) as well as the energy shift terms for single particle, pair and scattering processes in Eqs. (5.46, 5.57, 5.63) are zero,

## 5.6. QUANTUM MASTER EQUATION OF LINDBLAD TYPE

finally leads to the condensate quantum master equation of Lindbad type for a Bose-Einstein condensate of $N$ atoms:

$$\begin{aligned}\frac{\partial \hat{\rho}_0^{(N)}(t)}{\partial t} &= \sum_{\substack{N_0=0, \\ j=+,-}}^{N} \Gamma_N^j(N_0, T) \left[ \hat{\mathscr{S}}_j(N_0) \hat{\rho}_0^{(N)}(t) \hat{\mathscr{S}}_j^\dagger(N_0) - \frac{1}{2} \left\{ \hat{\mathscr{S}}_j^\dagger(N_0) \hat{\mathscr{S}}_j(N_0), \hat{\rho}_0^{(N)}(t) \right\}_+ \right] \\ &+ \sum_{\substack{N_0=0, \\ j=+,-}}^{N} \gamma_N^j(N_0, T) \left[ \hat{\mathscr{P}}_j(N_0) \hat{\rho}_0^{(N)}(t) \hat{\mathscr{P}}_j^\dagger(N_0) - \frac{1}{2} \left\{ \hat{\mathscr{P}}_j^\dagger(N_0) \hat{\mathscr{P}}_j(N_0), \hat{\rho}_0^{(N)}(t) \right\}_+ \right] .\end{aligned} \quad (5.68)$$

Remarkably, the dynamics of spatially coherent matter waves in the Bose gas below $T_c$ reflects itself in the Fock number representation as random, stochastic fluctuations of the condensate particle number described by a quantum Markov master equation of Lindblad type. The quantum jump operators $\hat{\mathscr{S}}_\pm(N_0)$ for single particle events, and $\hat{\mathscr{P}}_\pm(N_0)$ for pair events are defined via Eqs. (5.43, 5.44, 5.54, 5.55). We conclude this part of the thesis with the Lindblad master equation in Eq. (5.68).

# Part III

# Environment-induced dynamics in Bose-Einstein condensates

*Die Formel drückt also indirekt eine gewisse Hypothese über die gegenseitige Beeinflussung der Moleküle von vorläufig ganz rätselhafter Art aus, welche eben die gleiche statistische Wahrscheinlichkeit der hier als "Komplexionen" definierten Fälle bedingt. ...Man kann ihn auch beim Gase in entsprechender Weise deuten, indem man dem Gase in passender Weise einen Strahlungsvorgang zuordnet und dessen Interferenzschwankungen berechnet...*

Albert Einstein, 8. Januar 1925 [28]

Chapter 6

# Monitoring the Bose-Einstein phase transition

Bose-Einstein condensates are exquisite tools to study fundamental quantum phenomena on a micrometer scale. A vast range of different physical situations has been experimentally realized with ultracold matter in the last decade, confirming the fundamental importance and the broad applicational scope of Bose-Einstein condensation. However, a complete quantitative understanding of condensate formation remains one of the most striking theoretical topics of ultracold matter physics up to date. So far, the pioneering works [62, 89] were followed by quantitative theories [63, 76, 79], describing Bose-Einstein condensation in terms of average condensate growth. Hence, up to now, the dynamical inset of Bose-Einstein condensation is known to express as a spontaneously insetting exponential growth of the average condensate population after sudden cooling [90] of the gas below its critical temperature, see Fig. 6.3. The connection between the time evolution of the microscopic condensate number distribution during the Bose-Einstein phase transition relating the $N$-body dynamics in the Bose gas to the observation of an average macroscopic ground state occupation, however, is so far not well understood. A further step towards the latter aspect is approached by monitoring the condensate number fluctuations in Fock particle number representation during

Bose-Einstein condensation within our quantum master equation theory.

For this purpose, the master equation for the condensate particle number distribution is extracted from the Lindblad Eq. (5.68) in Section 6.1. It describes the Markov time evolution of the entire $N$-body state of the gas in terms of the quantum mechanical two body transition rates of Chapter 5. The quantum master equation serves as an ideal toy system to study the dynamics of critical number fluctuations during a quantum phase transition, enabling us to numerically study the full condensate quantum distribution during condensate formation in Section 6.2. The first direct access to the condensate and non-condensate number fluctuation dynamics during Bose-Einstein condensation provides a dynamical picture of the condensate growth resulting from the spatial and thermal averaging over all accessible single states of the gas particles below $T_c$, which manifests effectively in a randomly fluctuating, non-condensate environment, populating the ground state mode macroscopically on average. The average condensate growth and the non-condensate depletion garnished by the fluctuations of the non-condensate gas particles are analyzed in Sections 6.2.3, 6.3 and ??. Resulting initiation and formation times for Bose-Einstein condensation are compared to experimental observations [24, 91] and to previous theoretical predictions of quantum kinetic theory [17] in Sections 6.3 and ??.

## 6.1 Dynamical equations for Bose-Einstein condensation

Remember that, by Eq. (5.11),

$$p_N(N_0, t) = \langle N_0 | \text{Tr}_\perp \hat{\rho}^{(N)}(t) | N_0 \rangle \tag{6.1}$$

also specifies the total state of $N$ particles in the Bose gas:

$$\hat{\sigma}^{(N)}(t) = \sum_{N_0=0}^{N} p_N(N_0, t) | N_0 \rangle \langle N_0 | \otimes \hat{\rho}_\perp (N - N_0, T) \,, \tag{6.2}$$

# 6.1. DYNAMICAL EQUATIONS FOR BOSE-EINSTEIN CONDENSATION

see Eq. (5.11).

### 6.1.1 Master equation of Bose-Einstein condensation

The most general Lindblad quantum master equation (5.68) describes the time evolution of the reduced diagonal condensate density matrix, $\hat{\rho}_0(t) = \text{Tr}_\perp \hat{\sigma}^{(N)}(t)$, during Bose-Einstein condensation with respect to two body interactions in a gas of $N$ atoms. For practical purposes, it is useful to extract the master equation for the condensate number distribution in order to study the time evolution of the $N$-body state in Eq. (6.2).

To this end, the chemical potential of the condensate $\mu_0$ is considered to be energetically below non-condensate single particle energies, such as in experiments where condensation occurs on the single particle ground state level [15, 64, 92]. In that case, the Lindblad term for pair processes ($\rightsquigarrow\rightsquigarrow$) is negligible, and single particle processes ($\rightsquigarrow$) dominate the condensation process (see Section 7.1.2). Projecting the Lindblad master Eq. (5.68) onto the elements of the underlying condensate Fock-Hilbert space, the time evolution of the $N$-body state in Eq. (6.2) reduces to one closed master equation of Bose-Einstein condensation, which describes the condensation process as a consequence of single particle quantum jumps of the condensate particle number, $N_0 \to N_0 \pm 1$, induced by the non-condensate component of the gas:

$$\begin{aligned}\frac{\partial p_N(N_0, t)}{\partial t} =& -[\Gamma_N^+(N_0, T) + \Gamma_N^-(N_0, T)] p_N(N_0, t) \\ & + \Gamma_N^+(N_0 - 1, T) p_N(N_0 - 1, t) \\ & + \Gamma_N^-(N_0 + 1, T) p_N(N_0 + 1, t) \,, \end{aligned} \quad (6.3)$$

with the total condensate growth rate $\Gamma_N^+(N_0, T) = 2(N_0+1)\lambda_{\rightsquigarrow}^+(N-N_0, T)$, with $\lambda_{\rightsquigarrow}^+(N-N_0, T)$ defined by Eq. (5.40), and the total condensate loss rate $\Gamma_N^-(N_0, T) = 2N_0\lambda_{\rightsquigarrow}^-(N-N_0, T)$, with $\lambda_{\rightsquigarrow}^-(N-N_0, T)$ defined by Eq. (5.51). Remember that $p_N(N_0, t)$ also determines the non-condensate particle number distribution, because the total particle

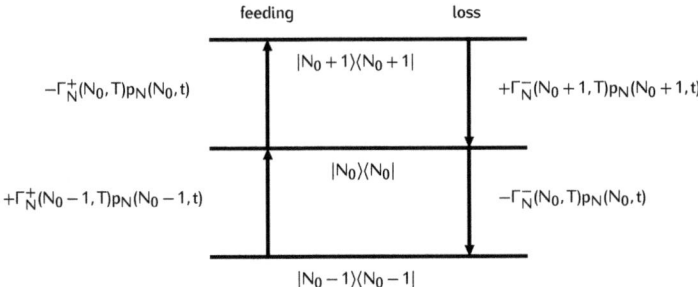

Figure 6.1: *Probability flow between different condensate number states, expressed by the master equation of Bose-Einstein condensation in Eq. (6.3). The corresponding transition rates are defined by $\Gamma_N^+(N_0,T) = 2(N_0+1)\lambda_{\leadsto}^+(N-N_0,T)$, with $\lambda_{\leadsto}^+(N-N_0,T)$ given by Eq. (5.40), and by $\Gamma_N^-(N_0,T) = 2N_0\lambda_{\leadsto}^-(N-N_0,T)$, with $\lambda_{\leadsto}^-(N-N_0,T)$ given by Eq. (5.51). In the stationary state, which is reached for long times $t \to \infty$, the rates obey the condition of detailed balance: $\Gamma_N^+(N_0,T)p_N(N_0,T) = \Gamma_N^-(N_0+1,T)p_N(N_0+1,T)$.*

number $N_0 + N_\perp = N$ is conserved, leading to $p_N(N - N_0, t) = p_N(N_0, t)$.

The master Eq. (6.3) needs to be distinguished from the so called Pauli master equation for the harmonic oscillator [20, 21] coupled to a heat bath, in which the transition rates are introduced phenomenologically by the Fermi's golden rule [20]. In contrast, Eq. (6.3) describes the condensate subsystem coupled to a non-condensate particle reservoir accounting for (i) the finite spatial phase coherence time ($\tau_{\text{col}}$) between the scattering quantum particles between system and reservoir leading to a finite resolution in energy $\Gamma \sim \tau_{\text{col}}^{-1}$ (see Chapter 5), as well as (ii) the finite size (particle number conservation implies $N - N_0$ particles for each condensate population of $N_0$ particles) of the particles in the non-condensate vapor below $T_c$, which leads to condensate formation. The steady state condensate quantum distribution is therefore not a thermal Boltzmann distribution over the eigenenergies of the harmonic oscillator [21, 93] as governed by thermal atoms, but predicts macroscopic ground state occupation below $T_c$.

The resulting probability flow (condensate growth) in particle number repre-

## 6.1. DYNAMICAL EQUATIONS FOR BOSE-EINSTEIN CONDENSATION

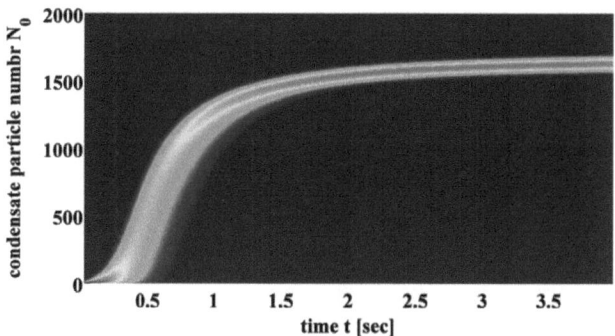

Figure 6.2: *Time evolution of $p_N(N_0, t)$ with respect to Eq. (6.3) (low (blue) and high (red) intensity regions, indicated by the color gradient) during the transition of $N = 2000$ $^{87}Rb$ atoms into a Bose-Einstein condensate in a harmonic trap with frequencies $\omega_x = \omega_y = 2\pi \times 42.0$ Hz, $\omega_z = 2\pi \times 120.0$ Hz. The final temperature of the gas is $T = 20.31$ nK; the critical temperature is $T_c = 33.86$ nK. Whereas the wave picture of the particles implies the interference of all waves below $T_c$ to a giant matter wave (see Chapter 1), condensate formation translates in the many particle picture as a rapid growth of the average condensate fraction, garnished with large initial condensate number fluctuations, as discussed in more detail in Section 6.2.*

sentation is sketched in Fig. 6.1: the net particle flow *towards* a state $|N_0\rangle\langle N_0| \otimes \hat{\rho}_\perp(N - N_0, T)$ is due to the terms of the positive probability feeding current $\Gamma_N^+(N_0 - 1, T)p_N(N_0 - 1, t) + \Gamma_N^-(N_0 + 1, T)p_N(N_0 + 1, t)$, whereas the particle flow *from* the state $|N_0\rangle\langle N_0| \otimes \hat{\rho}_\perp(N - N_0, T)$ is governed by the negative probability loss current $\Gamma_N^+(N_0, T)p_N(N_0, t) + \Gamma_N^-(N_0, T)p_N(N_0, t)$. As shown in Section 8.1, the steady state of the system is thus reached, if, and only if the net probability flux to every number state $|N_0\rangle\langle N_0|$ (implying in particular detailed balanced particle flow, $\partial_t \langle N_0 \rangle = 0$) is zero, i.e. $\Gamma_N^+(N_0, T)p_N(N_0, T) = \Gamma_N^-(N_0 + 1, T)p_N(N_0 + 1, T)$ for all $N_0$.

To study the dynamics of condensate formation, we solve the $(N + 1)$ coupled differential Eqs. (6.3) for the condensate particle number distribution $p_N(N_0, t)$ by numerically exact propagation, using the $2(N+1)$ feeding and loss rates $\lambda_\pm^\pm(N-N_0, T)$ in Eq. (7.33). They particularly define the time evolution of the average condensate

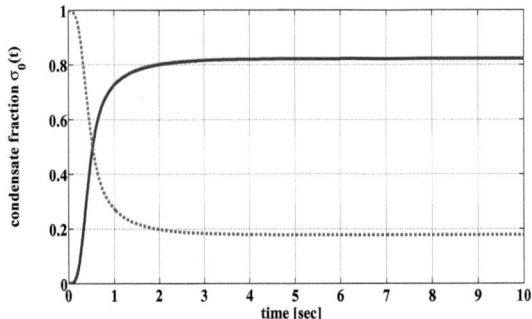

Figure 6.3: *Average condensate fraction $\sigma_0(t) = \langle N_0 \rangle(t)/N$ (blue solid line) and non-condensate fraction $\sigma_\perp(t) = 1 - \sigma_0(t)$ (red dashed line) during Bose-Einstein condensation in a gas of $N = 2500$ $^{87}$Rb atoms obtained from Eqs. (6.3, 6.4). Trap parameters and final temperature correspond to Fig. (6.2). The critical temperature is $T_c = 36.47$ nK.*

occupation, $\langle N_0 \rangle(t)$, see Eq. (6.4). A typical example of the time evolution for $p_N(N_0, t)$ during Bose-Einstein condensation in a gas of $N = 2000$ $^{87}$Rb atoms is displayed in Fig. 6.2. In general, our numerical calculations in this parameter regime require small computation times of 10 – 60 seconds. For increasing total particle numbers, the scenario in Fig. 6.2 is reproducible for up to $N = 10^5 - 10^6$ atoms within two days of computation time on a single processor.

### 6.1.2 Growth equations for average condensate occupation

From Eq. (6.3), any desired moment of $p_N(N_0, t)$ can be extracted. Taking the average of $N_0$ over the probability distribution $p_N(N_0, t)$ leads to the growth of the average condensate fraction:

$$\langle N_0 \rangle(t) \equiv \sum_{N_0=0}^{N} N_0 p_N(N_0, t) \ . \tag{6.4}$$

## 6.1. DYNAMICAL EQUATIONS FOR BOSE-EINSTEIN CONDENSATION

Figure 6.3 shows the average ground state occupation during Bose-Einstein condensation for the same same trap parameters as in Fig. 6.2, with $N = 2500$ $^{87}$Rb atoms undergoing the Bose-Einstein phase transition to the final gas temperature $T = 20.0$ nK. The typical S-shape behavior [59, 18] of the condensate fraction $\sigma_0(t) = \langle N_0 \rangle(t)/N$ is confirmed by our quantum master equation (6.3), such as the inverse Z-shape behavior for the non-condensate fraction, $\sigma_\perp(t) = 1 - \langle N_0 \rangle(t)/N$, is presented as a function of time.

Instead of propagating the exact equation (6.4) for calculating the average condensate occupation, it is possible to deduce a simple growth equation, similar to quantum kinetic theory [76]. Herefore, the quantum master Eq. (6.3) is traced over the number of condensate particles, $N_0$, thereby leading to

$$\frac{\partial \langle N_0 \rangle(t)}{\partial t} = \sum_{N_0=0}^{N} 2(N_0+1)[\lambda_{\leadsto}^+(N-N_0, T)p_N(N_0, t) - \lambda_{\leadsto}^-(N-N_0-1, T)p_N(N_0+1, t)] \ . \quad (6.5)$$

Rather than completely neglecting quantum fluctuations [76] (the width of $p_N(N_0, t)$), we consider a sufficiently narrow peaked distribution $p_N(N_0, t)$ around the mean value $\langle N_0 \rangle$, within which the rates are approximately constant (see Figs. 6.2, 7.6), $\lambda_{\leadsto}^\pm(N-N_0, T) \approx \lambda_{\leadsto}^\pm(N-\langle N_0\rangle, T)$, for $N_0 \simeq \langle N_0 \rangle$. This entails the condensate growth equation for the average condensate occupation gas of exactly $N$ particles:

$$\frac{\partial \langle N_0 \rangle(t)}{\partial t} = 2[\lambda_{\leadsto}^+(N-\langle N_0\rangle, T) - \lambda_{\leadsto}^-(N-\langle N_0\rangle, T)](\langle N_0\rangle(t)+1) \ . \quad (6.6)$$

Equation (6.6) nicely highlights the quantum coherent nature of the condensation process: The net flux rate $\lambda_{\leadsto}^+(N-\langle N_0\rangle, T) - \lambda_{\leadsto}^-(N-\langle N_0\rangle, T)$ per particle towards the condensate mode (the ratio gives the balance of the number of events in which the particle populates the condensate mode to the number of events it entered the non-condensate) is stimulated by the factor $(N_0+1)$, meaning that the presence of a condensate enhances the net feeding rate of each individual non-condensate particle.

Equation (6.6) differs — despite the transition rates which depend on the non-condensate particle number $N - N_0$, and are obtained without further approximations than assuming the dilute gas limit (compare the rates of QKT in Section 2.4) — in the spontaneous emission term $2\lambda_\leftrightarrows^+(N-\langle N_0\rangle, T)$ from the kinetic growth equation of QKT in Eq. (2.31). The absence of the emission term in Eq. (6.6) is conceptually important, reflecting consistency with thermodynamics: for $\partial_t \langle N_0 \rangle = 0$, the net energy flow between condensate and non-condensate is zero on average, $\mu_0 = \mu_\perp$, which is easily verified using the balance condition, $\lambda_\leftrightarrows^+(N-\langle N_0\rangle, T) = \exp[\beta \Delta \mu (N-\langle N_0\rangle, T)]\lambda_\leftrightarrows^-(N-\langle N_0\rangle, T)$, in Eq. (7.20), with $\Delta \mu(N-\langle N_0\rangle, T) = \mu_\perp(N-N_0, T) - \mu_0$. The condensate growth scenario in Eq. (6.6) thus implies that $\mu_0 = \mu_\perp(N-\langle N_0\rangle, T)$ on average at stationary particle flow between condensate and non-condensate. The latter is in particular reached in the steady state of the gas (see Chapter 8) — in agreement with thermodynamics. According to the validity of the detailed balance condition, an upper bound for deviations from microscopic energy conservation is $\beta \hbar \tau_{\text{col}}^{-1} \ll 1$ (see Chapter 7). The modified Eq. (6.6), however, yields only negligibly small quantitative corrections (see Chapter 6.3) with respect to Eq. (2.31).

### 6.1.3 Condensate particle number fluctuations

Condensate number fluctuations during Bose-Einstein condensation are characterized by the second moment of $p_N(N_0, t)$:

$$\text{Var}[N_0](t) = \langle N_0^2 \rangle - \langle N_0 \rangle^2 = \sum_{N_0=0}^{N} N_0^2 p_N(N_0, t) - \left[ \sum_{N_0=0}^{N} N_0 p_N(N_0, t) \right]^2 . \quad (6.7)$$

From the evolution Eq. (6.3), the variance $\text{Var}[N_0](t)$, and thus the standard deviation $\Delta N_0(t) = \sqrt{\text{Var}[N_0](t)}$ of the distribution $p_N(N_0, t)$ is extracted as a function of time. The standard deviation of the condensate number $\Delta N_0(t)$ is displayed in Fig. 6.4 during Bose-Einstein condensation, using the same parameters as in Fig. 6.3.

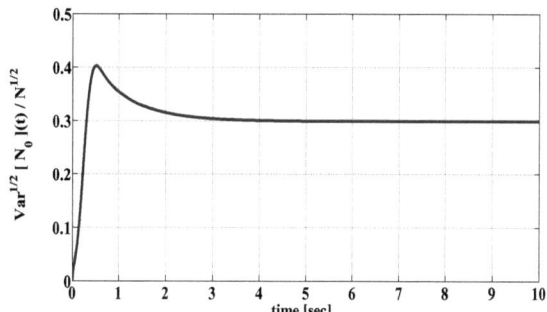

Figure 6.4: *Standard deviation $\Delta N_0(t)$ of the condensate particle number distribution $p_N(N_0, t)$ during Bose-Einstein condensation for the same parameters as in Fig. 6.3. Large number fluctuations in the non-condensate thermal vapor are observed at the initial stage of formation with a maximum spread of the distribution at $t \approx 600$ ms, reducing at stationary condensate growth, see Fig. 6.3.*

## 6.2 Bose-Einstein condensation in harmonic traps

To analyze the Bose-Einstein phase transition in more detail, we consider here a gas of $N$ interacting particles (in the perturbative limit, $a\varrho^{1/3} \to 0^+$, see Chapter 7) prepared in a time-independent three-dimensional harmonic trapping potential with frequencies $\omega_x, \omega_y$ and $\omega_z$, in an initial mixed many particle quantum state

$$\hat{\sigma}^{(N)}(0) = \sum_{N_0=0}^{N} p_N(N_0, 0)|N_0\rangle\langle N_0| \otimes \hat{\rho}_\perp(N - N_0, T) \,, \tag{6.8}$$

given the initial condensate particle number distribution $p_N(N_0, 0) = e^{-\beta\eta_0 N_0}(1 - e^{-\beta\eta_0(N+1)})$ (of a gas above $T_c$) corresponding to thermal atoms [93]. Now, the relaxation dynamics of this initial state to the Bose condensed phase is studied, directly reflecting the dynamical onset of non-condensate quantum fluctuations in terms of the condensate and non-condensate particle number distribution $p_N(N_0, t)$

as described by Eq. (6.3).

### 6.2.1 Monitoring of the condensate number distribution

The formation of a Bose-Einstein condensate is studied within our master equation theory by direct numerical propagation of Eq. (6.3), leading to a system of $N+1$ coupled differential equations for each instant of time. Numerical solutions have been reproduced[1] for different experimental parameters and particle numbers up to $N = 5.0 \cdot 10^5$.

A typical example is displayed in Fig. 6.5: The probability distribution $p_N(N_0, t)$ is shown as a function of time $t$ and of the condensate particle number $N_0$, for a gas of $N = 200$ $^{87}$Rb atoms which undergoes the Bose-Einstein phase transition in a three-dimensional harmonic trap with frequencies $\omega_x = \omega_y = 2\pi \times 42.0$ Hz, $\omega_z = 2\pi \times 120.0$ Hz. The final gas temperature is set to $T/T_c = 0.40$ in order to model a sudden cooling process [90], given the ideal gas critical temperature $T_c = 15.72$ nK. The lower panel in Fig. 6.5 shows the x-y projection of the condensate particle number distribution $p_N(N_0, t)$.

A switch below the critical temperature induces a quench[2] of the non-condensate density below the critical density for Bose-Einstein condensation, $(N-N_0)\hbar^3\omega_x\omega_y\omega_z /(k_B T)^3 > \zeta(3)$, and thus leads to a coherent spatial coupling of the atoms in the gas. While the wave picture of the particles implies an initially large number of coherently interfering matter waves, this wave dynamics is translated into the microscopic many particle picture as a coupling of many transition channels for particle exchanges between non-condensate and condensate atoms leading to number uncertainties. The resulting initially strong number fluctuations during the onset condensate of the condensate formation process in the gas initiate and trigger the Bose-Einstein phase transition, as directly monitored in Fig. 6.5: In the exponential step of condensate growth, a large initial spreading of the condensate particle

---
[1]with computation times of up to 1 — 2 days on a serial computer
[2]Consult Chapter 7 and Section 7.6 for the conditions of the critical density for condensate formation, as well as for the reaching of a detailed particle balance.

## 6.2. BOSE-EINSTEIN CONDENSATION IN HARMONIC TRAPS

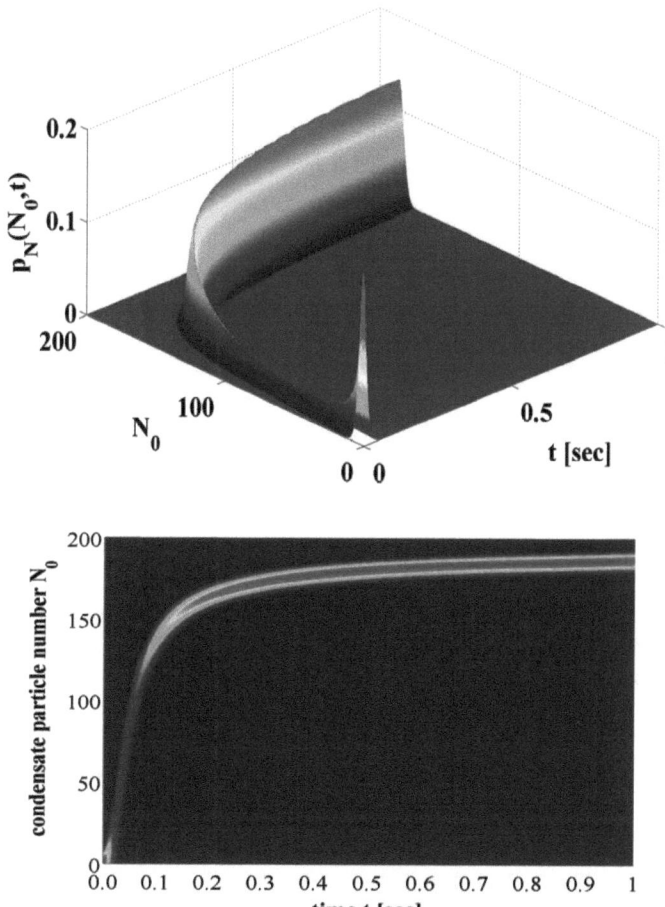

Figure 6.5: *Condensate particle number distribution $p_N(N_0, t)$ during the transition of a gas of $N = 200$ $^{87}$Rb atoms into a Bose-Einstein condensate in a three-dimensional harmonic trap with trapping frequencies $\omega_x = \omega_y = 42.0$ Hz, $\omega_z = 120.0$ Hz, given a critical temperature $T_c = 15.72$ nK. We model a sudden cooling process by switching the temperature of the reservoir below the phase transition temperature, $T/T_c = 0.4$. Note that the non-condensate particle number distribution, $p_N(N - N_0, t) = p_N(N_0, t)$, is simultaneously captured by replacing $N_0 \rightarrow (N - N_0)$ in the figure. Lower panel shows the x-y projection of $p_N(N_0, t)$.*

number distribution $p_N(N_0, t)$ is observed, which is due to the large number fluctuations in the non-condensate environment being transferred to the condensate subsystem because of particle number conservation. The corresponding number distributions are both simultaneously captured by $p(N_0, t) = p_N(N - N_0, t)$. In this first stage, the buildup of a large condensate fraction sets in, however, with high initial uncertainty in the number occupation.

As the gas evolves in time, the non-condensate density reduces and approaches its equilibrium value defined by the final temperature $T$ of the gas, the non-condensate atoms increasingly lose their spatial coherence during condensate growth (in the wave picture), and thus finally stop the condensation process. In the many particle representation, this is effectively reflected by the depopulation of non-condensate single particle occupations, and consequently by the reduction of the condensate number fluctuations. The reshaping of the condensate particle number distribution is herein understood as due to the decreasing number of contributing non-condensate single particle states to the dynamical condensate growth.

In the final steady state, the remaining condensate number fluctuations are induced by thermal fluctuations of the surrounding non-condensate environment (see Chapter 8).

It will be shown in Chapter 7 that, on average, the reaching of a detailed balanced particle flow between condensate and non-condensate modes implies a (dilute) equilibrium non-condensate density, i.e. the stationarity condition $\varrho_\perp \lambda^3(T) = \zeta(3/2)$ for a gas in a box, or $(N - N_0)(k_B T)^3/\hbar^3 \omega_x \omega_y \omega_z = \zeta(3)$ for a three-dimensional harmonic potential (see Section 7.6). In the steady state, the non-condensate chemical potential is therefore zero (in both cases), meaning that the entropy of the gas is maximized, that the free energy is minimized, and that the net average energy flow between condensate and non-condensate is zero — according to the laws of thermodynamics. Obviously, the average ground state occupation (condensate formation) is enhanced below $T_c$ with respect to the initial thermal condensate number distribution $p_N(N_0, T) = e^{-\beta \eta_0 N_0}(1 - e^{-\beta \eta_0 (N+1)})$.

## 6.2. BOSE-EINSTEIN CONDENSATION IN HARMONIC TRAPS

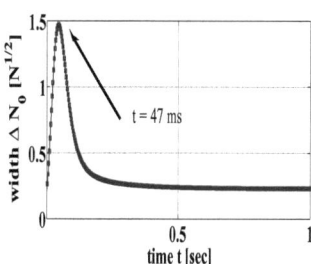
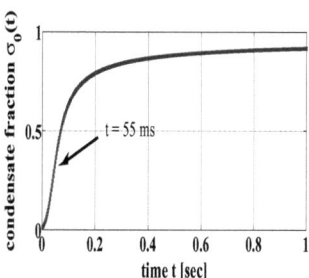

Figure 6.6: *Time dependent width, $\Delta N_0(t) = [\langle N_0^2 \rangle(t) - \langle N_0 \rangle^2(t)]^{1/2}$ (left panel), and the evolution of the maximum $\sigma_0(t) = \langle N_0 \rangle(t)/N$ (right panel), of the distribution $p_N(N_0, t)$ in Fig. 6.5 during Bose-Einstein condensation. The inflection point occurs of the with occurs at an approximate condensate fraction of 1/4.*

### 6.2.2 Dynamics of the condensate number variance

In Figs. 6.6 and 6.7, the dynamical three-step process of the number distribution $p_N(N_0, t)$ in Fig. 6.5 is emphasized by extracting the time evolution of the distribution's maximum,[3] $\sigma_0(t) = \langle N_0 \rangle/N$ (right panel), and its width $\Delta N_0(t)/\sqrt{N} = [\langle N_0^2 \rangle(t) - \langle N_0 \rangle^2(t)]^{1/2}/\sqrt{N}$ (standard deviation, left panel), as a function of time: (i) In the exponential stage of condensate growth, the condensate particle number starts to fluctuate, before (ii) its distribution is reshaped at the inflection point $\langle N_0 \rangle/\langle N_\perp \rangle \sim 1/3$. After the initial cycle, (iii) the exponential growth stops and the particle number distribution tends towards its final equilibrium shape of a well-defined width.

The maximum width (inflection point) occurs at $t = 47$ ms, i.e. at an average condensate fraction of $\sigma_0 = 0.25$, and reaches the equilibrium width of the particle number distribution $p_N(N_0, T)$ approximately after $t \sim 500$ ms. The observed inflection point turns out to be universal (at $\sigma_0 \sim 0.25$) in the numerical propagation of Eq. (6.3). This can be understood as due to a direct consequence of

---
[3] in units of the total particle number $N$

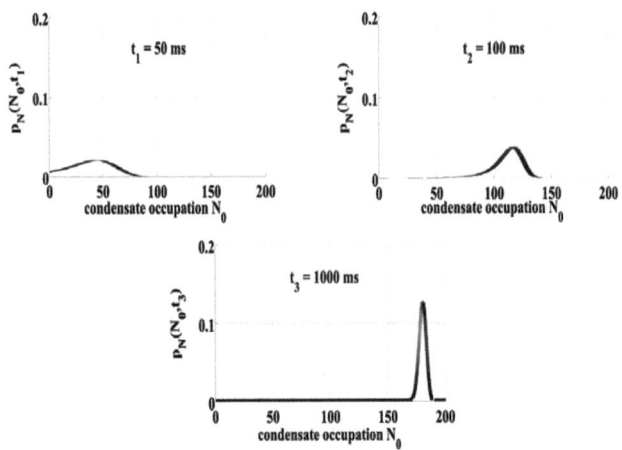

Figure 6.7: *Sequences of characteristic dynamical stages of the condensate particle number distribution $p_N(N_0,t)$ during condensate formation, here after $t_1 = 50$ ms, $t_2 = 100$ ms and $t_3 = 1000$ ms, extracted from Fig. 6.5. The incipiently narrow distribution is largely spread at the stage of exponential condensate growth, and begins to evolve towards high condensate population in the second stage. Finally, it reshapes while reaching the equilibrium steady state distribution in the last stage of (linear) condensate growth.*

the specific form of the interaction term $\hat{V}_{\leadsto}$ in Eq. (4.16), because the approximate ratio of $\langle N_0 \rangle / \langle N_\perp \rangle = 1/3$ corresponds to interactions between condensate and non-condensate atoms getting maximal with regard to single particle processes.

Analysis of the average condensate fraction, $\sigma_0(t) = \langle N_0 \rangle / N$ in Fig. 6.6, highlights that the inflection point at the exponential stage of condensate growth is slightly delayed with respect to the widths' time evolution, occuring at $t = 55$ ms. The condensate fraction reaches its equilibrium value starting around $t = 1000$ ms, about a factor of 2 later than the width $\Delta N_0(t)$. This indicates that the rapid initial depopulation of the highly excited single particle states reduce the condensate number fluctuations faster than reaching a steady state particle flow between the energet-

## 6.2. BOSE-EINSTEIN CONDENSATION IN HARMONIC TRAPS

ically low-lying non-condensate single particle states and the condensate mode, being not yet the case after 500 ms (comparable to Stoof's prediction explained in Chapter 2.4).

In Fig. 6.7, three different sequences of the distribution $p_N(N_0, t)$ are extracted from Fig. 6.6 at $t_1 = 50$ ms, $t_2 = 100$ ms and $t_3 = 1000$ ms, to highlight the three characteristics steps of the distribution $p_N(N_0, t)$ towards the stationary state.

### 6.2.3 Average condensate growth from the thermal cloud

To study the dynamics of condensate growth and non-condensate depletion in the microscopic many particle picture, we have used so far the exact numerical propagation of Eq. (6.3), taking into account the full number distribution (see Chapter 6). Here, we analyze and employ the growth Eq. (6.6) in order to study condensate formation on a thermodynamic scale.

Equation (6.6) describes the process of condensate formation from an initially empty condensate mode, starting to grow proportional to the rate $\lambda_+^{\leftrightarrow}(N-\langle N_0 \rangle = 0, T)$, because initially $\langle N_0 \rangle = 0$, and therewith $\lambda_+^{\leftrightarrow}(N-\langle N_0 \rangle = 0, T) - \lambda_-^{\leftrightarrow}(N-\langle N_0 \rangle = 0, T) > 0$ (i.e., particle flow towards the condensate, see Section 7 for explicit calculations of the transition rates). A temperature switch below the critical temperature thus induces energy flow towards to condensate in terms of particles, $\Delta\mu(N-\langle N_0 \rangle, T) < 0$, which gives rise to an exponential condensate growth. The growth is stimulated by the factor $(\langle N_0 \rangle(t) + 1)$ after initiation which modifies and redefines the condensate feeding and loss rates $\lambda_\pm^{\leftrightarrow}(N-\langle N_0 \rangle, T)$ in time. This "loop" runs until stationarity is reached, where $\Delta\mu(N-\langle N_0 \rangle, T) \to 0^-$. In this case, detailed balanced particle flow $\lambda_+^{\leftrightarrow}(N-\langle N_0 \rangle, T) = \lambda_-^{\leftrightarrow}(N-\langle N_0 \rangle, T)$ and a macroscopic ground state occupation is reached on average.

It is the equal energy balance condition $\mu_0 = \mu_\perp(N - \langle N_0 \rangle)$, which defines the mean equilibrium occupations $\langle N_0 \rangle(\infty)$ and $\langle N_\perp \rangle(\infty) = 1 - \langle N_0 \rangle(\infty)$ by Eq. (7.8), quantitatively determined by the final gas temperature $T$, as well as by the geometry of the trap and the total number of particles $N$ in the system. Equilibrium occupa-

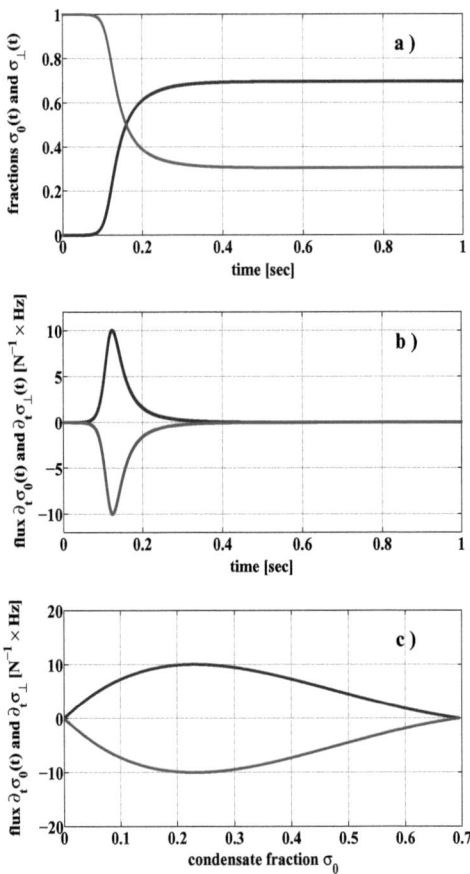

Figure 6.8: *a) Average condensate (blue line) and non-condensate (red line) fractions, $\sigma_0(t)$ and $\sigma_\perp(t)$, respectively, as a function of time for a gas of $N = 5 \cdot 10^5$ $^{23}$Na atoms undergoing the Bose-Einstein phase transition in a three-dimensional harmonic trap with frequencies $\omega_x = 2\pi \times 235$ Hz, $\omega_y = 2\pi \times 410$ Hz, $\omega_z = 2\pi \times 745$ Hz. The final temperature is $T = 1.0$ μK, and the ideal gas critical temperature is $T_c = 1.5$ μK. Parameters correspond to experiments of Ref. [24], with a gas parameter of $a\varrho^{1/3} \sim 10^{-2}$. Figures b) and c) show the temporal change of the average condensate fraction $\dot{\sigma}_0(t)$ (blue line), and non-condensate fraction $\dot{\sigma}_\perp(t)$ (red line), versus time in figure b), and versus condensate fraction in figure c), respectively.*

## 6.2. BOSE-EINSTEIN CONDENSATION IN HARMONIC TRAPS

tions $\langle N_0 \rangle(\infty)$ and $\langle N_\perp \rangle(\infty)$ are given analytically in the semiclassical limit of large particle numbers and high temperature via Eq. (7.48): For the three-dimensional uniform case, the corresponding equilibrium values are $\langle N_0 \rangle(\infty) = N(1 - T^{3/2}/T_c^{3/2})$ with $T_c = 2\pi\hbar^2 \varrho^{2/3}/\zeta^{2/3}(3/2) m k_B$, whereas $\langle N_0 \rangle(\infty) = N(1 - T^3/T_c^3)$ with $T_c = \hbar\bar{\omega} N^{1/3}/\zeta^{1/3}(3) k_B$ for a three-dimensional harmonic trap, where $\bar{\omega} = (\omega_x \omega_y \omega_z)^{1/3}$ is the spatially averaged trap frequency (see Chapter 7). Corrections to the critical temperature (arising from the discrete nature of the single particle spectrum) are discussed in Chapter 8.

Figure 6.8a displays the dynamics of the average condensate fraction $\sigma_0(t) = \langle N_0 \rangle(t)/N$ and of the average non-condensate fraction $\sigma_\perp(t) = \langle N_\perp \rangle(t)/N$, correspondingly, as a function of time, for a condensate formation process with typical experimental parameters [24]. Similarly, $N = 5 \cdot 10^5$ $^{23}$Na atoms are implemented to study the formation process in a three-dimensional harmonic trap with frequencies $\omega_x = 2\pi \times 235$ Hz, $\omega_y = 2\pi \times 410$ Hz, $\omega_z = 2\pi \times 745$ Hz. The gas is subjected to the initial conditions $\sigma_0(0) = 0.0$ and $\sigma_\perp(0) = 1.0$, and $T = 1.0$ $\mu$K. The critical temperature is $T_c = 1.5$ $\mu$K. According to Ref. [24] the dilute gas parameter is $a\varrho^{1/3} \sim 10^{-2} \ll 1$.

We confirm[4] the typical "S-shape behavior" of Bose-Einstein condensation [17, 67, 72, 73, 74, 76]: After a short initiation time, the average condensate fraction (depicted as blue line) grows exponentially fast, before it slowly reaches its equilibrium population. Since the particle number is conserved, the inverse scenario is observed for the non-condensate fraction (shown as red line), i.e., after a slow initiation period, the non-condensate starts to exponentially decrease until reaching the equilibrium steady state. It is evident that the sum $\sigma_0(t) + \sigma_\perp(t) = 1$ at all times[5].

Figures 6.8b and 6.8c show the average flux $\dot{\sigma}_0(t)$ of particles from the non-condensate to the condensate, and vice versa, with $\dot{\sigma}_\perp(t) = -\dot{\sigma}_0(t)$, as a function of time (Fig. 6.8b), and as a function of the condensate fraction (Fig. 6.8c), respectively, for the same parameters as in Fig. 6.8a. For the given parameters the initiation

---
[4]as being analyzed for large ranges of different experimental parameters
[5]Indeed, Eq. (6.3) analytically shows that $p_N(N_0, t)$ remains normalized at all times. In consequence, $N_0 + N_\perp = N$ maintains not only on average but also for each realization, because $p(N_0, t) \equiv p_N(N - N_0, t)$.

time[6] is $\tau_{ini} = 125$ ms, and the final saturation time to the equilibrium population is $\tau_0 = 400$ ms. The observed time scales for condensate formation hence agree with the theoretical quantitative analysis presented in Refs. [91] for sudden cooling, and thus with the experimental observations of Refs. [24], see the following section.

## 6.3 Comparison of formation times to state-of-the-art

Quantitative comparisons of initiation and condensate formation times predicted by the master equation theory (based on the unperturbed one-body condensate feeding rates $\lambda_{\sim}^{\pm}(N-N_0, T)$ of Section 7.5) are compared to experimental and theoretical predictions of Refs. [24, 17, 91]. To this end, we use the kinetic growth equation in Eq. (6.6).

The initiation time $\tau_{ini} = 125$ ms in Fig. 6.8 is in very good agreement with QKT [91], where the authors found an initiation time $\tau_{ini} = 120 - 130$ ms, and thus with the experiments in Ref. [24]. However, the condensate formation time $\tau_0 \sim 500$ ms predicted by our master equation approach deviates by a factor two from the formation time $\tau_0 \sim 250$ ms of quantum kinetic theory [91]. Our condensate formation times match the correct order of magnitude of the experimental setup and previous theoretical predictions in this case.

Fig. 6.9 presents a further comparison to experimental and theoretical results [17] in case of sudden cooling. To model these experiments, we consider a cloud of $N = 4.4 \cdot 10^6$ bosonic $^{87}$Rb atoms, cooled in an anisotropic harmonic trapping potential with trapping frequencies $\omega_x = \omega_y = 2\pi \times 110$ Hz, $\omega_z = 2\pi \times 14$ Hz. Since the final temperature of the atomic cloud in the experiment is $T = 220 \pm 20$ nK, the condensate formation time is calculated for three different temperatures $T = 200$ nK, $T = 220$ nK and $T = 240$ nK, see Fig. 6.9: Initiation times are $\tau_{ini} = 106$ ms, $\tau_{ini} = 243$ ms and $\tau_{ini} = 522$ ms, strongly dependent on the final gas temperature. Formation times range from $\tau_0 = 800$ to $\tau_0 = 1500$ ms. Initiation times of the experiment [17] are

---

[6]The initiation time is defined as the point of maximum particle flux to the condensate mode, specified by the condition that $\partial^2 \langle N_0 \rangle(t)/\partial t^2 = 0$.

## 6.3. COMPARISON OF FORMATION TIMES TO STATE-OF-THE-ART

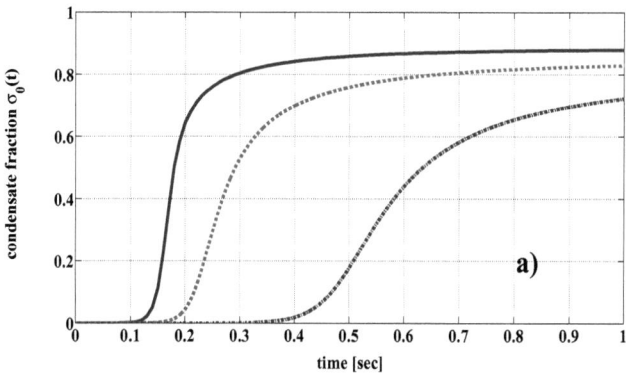

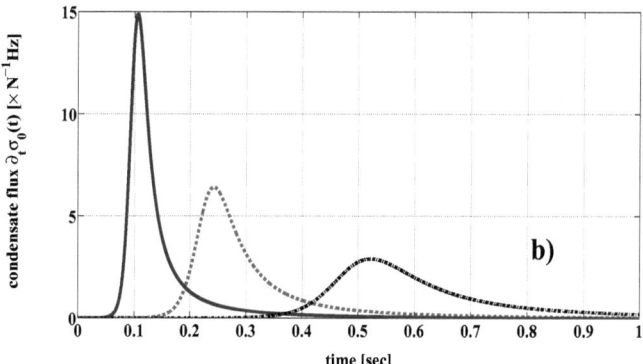

Figure 6.9: *Average condensate fraction $\sigma_0(t)$ in a), and average particle flux $\dot{\sigma}_0(t)$ to the condensate in b), for the experimental parameters [17] corresponding to $N = 4.4 \cdot 10^6$ bosonic $^{87}$Rb atoms in an anisotropic harmonic trapping potential with trapping frequencies $\omega_x = \omega_y = 2\pi \times 110$ Hz, $\omega_z = 2\pi \times 14$ Hz, for final gas temperatures $T = 200$ nK (blue solid line), $T = 220$ nK (red dashed line) and $T = 240$ nK (black dashed dotted line). We find initiation times $\tau_{ini} = 105$ ms for $T = 200$ nK, $\tau_{ini} = 243$ ms for $T = 220$ nK, and $\tau_{ini} = 522$ ms for $T = 240$ nK, and condensate formation times, which range from $\tau_0 = 800$ to $\tau_0 = 1500$ ms. The time scales of Figs. a) and b) match the experimental ones of Ref. [17] within the experimental accuracy.*

$\tau_{\text{ini}} \sim 300$ ms, while condensate formation times are of the order of $\tau_0 \sim 800-900$ ms.

A similar quantitative agreement between our master equation theory and the experiment has been found for further experimental measurements [23].

We conclude that condensate formation times obtained in the perturbative limit, $a\varrho^{1/3} \to 0^+$, match realistic experimental time scales for condensate formation [17, 24, 91]. Inclusion of the perturbative effects onto the condensate wave function is expected to provide even better quantitative agreement.

Chapter 7

# Transiton rates for Bose-Einstein condensation

We show here how to evaluate and analyze the corresponding transition rates and energy shifts formally introduced in Chapter 5 to solve the master equation for the diagonal elements of the condensate density matrix (6.3) arising from the Lindblad master equation (5.68). Explicit analytical expressions are given for the transition rates and energy shifts in a three-dimensonal harmonic trap.

## 7.1 Single particle ($\leadsto$), pair ($\leftrightsquigarrow$) and scattering ($\circlearrowleft$) rates

In this section, explicit analytical expressions for all two body transition rates for particle flow between the condensate and non-condensate are derived. These were formally specified as single particle condensate feeding and loss rates, $\lambda^{\pm}_{\leadsto}(N-N_0, T)$ in Eqs. (5.40, 5.41), as pair condensate feeding and loss rates, $\lambda^{\pm}_{\leftrightsquigarrow}(N-N_0, T)$ in Eqs. (5.51, 5.52), and as the scattering rate $\lambda_{\circlearrowleft}(N-N_0, T)$ in Eqs. (5.61, 5.62).

### 7.1.1 Single particle feeding and loss rate

Given a number of $N_0$ particles occupying the condensate mode and a final temperature $T$ of the gas, particle exchanges between condensate and non-condensate which raise and lower the condensate particle number by one occur with the rates $\lambda^{\pm}_{\leftrightsquigarrow}(N-N_0,T)$ in Eqs. (5.40, 5.41). To calculate the transition rates associated to single particle events, the lower bound of the time integral in the single particle feeding and loss rates in Eq. (5.40, 5.41) is extended to $-\infty$, using the property

$$\mathscr{G}^{(\pm)}_{\leftrightsquigarrow}(\vec{r},\vec{r}\,',N-N_0,T,\tau) = \left[\mathscr{G}^{(\pm)}_{\leftrightsquigarrow}(\vec{r},\vec{r}\,',N-N_0,T,-\tau)\right]^* \qquad (7.1)$$

of the two point correlation functions $\mathscr{G}^{(\pm)}_{\leftrightsquigarrow}(\vec{r},\vec{r}\,',N-N_0,T,\tau)$ for single particle processes in Eqs. (5.37, 5.38). Thereby, the single particle loss and feeding rates turn into

$$\lambda^{\pm}_{\leftrightsquigarrow}(N-N_0,T) = \frac{g^2}{2\hbar^2}\iint_{\mathscr{C}\times\mathscr{C}} d\vec{r}\,d\vec{r}\,'\, \Psi_0^*(\vec{r})\Psi_0(\vec{r}\,')\int_{-\infty}^{\infty} d\tau\, e^{\pm i\omega_0\tau - \Gamma^2\tau^2}\mathscr{G}^{(\pm)}_{\leftrightsquigarrow}(\vec{r},\vec{r}\,',N-N_0,T,\tau)\,. \qquad (7.2)$$

First, attention is drawn to the decomposition of the two point correlation function $\mathscr{G}^{(\pm)}_{\leftrightsquigarrow}(\vec{r},\vec{r}\,',N-N_0,T,\tau)$. As discussed in Section 4.3, the average of the non-condensate field correlations in $\mathscr{G}^{(\pm)}_{\leftrightsquigarrow}(\vec{r},\vec{r}\,',N-N_0,T,\tau)$ is evaluated with respect to a thermal state of the non-condensate described by the Hamiltonian $\hat{\mathcal{H}}_{\perp}$ in Eq. (4.26) projected onto the subspace of $(N-N_0)$ particles. According to Wick's theorem [94], which can generally be applied in order to calculate expectation values of field operators with respect to the linearized non-condensate at thermal equilibrium, the two point correlation functions $\mathscr{G}^{(\pm)}_{\leftrightsquigarrow}(\vec{r},\vec{r}\,',N-N_0,T,\tau)$ decompose from a product of six non-condensate fields, see Eq. (5.37, 5.38), into a product of three time ordered two point correlation functions of two non-condensate fields, as explicitly shown in Eqs. (A.9, A.10) of Appendix A.1. This leads to the expression

$$\lambda^{\pm}_{\leftrightsquigarrow}(N-N_0,T) = \frac{8\pi^3\hbar^2 a^2}{m^2}\sum_{k,l,m\neq 0}\mathscr{F}^{\pm}_{\leftrightsquigarrow}(k,l,m,N-N_0,T)\,\delta^{(\Gamma)}(\omega_k+\omega_l-\omega_m-\omega_0) \qquad (7.3)$$

## 7.1. SINGLE PARTICLE (~), PAIR (~~) AND SCATTERING (○) RATES

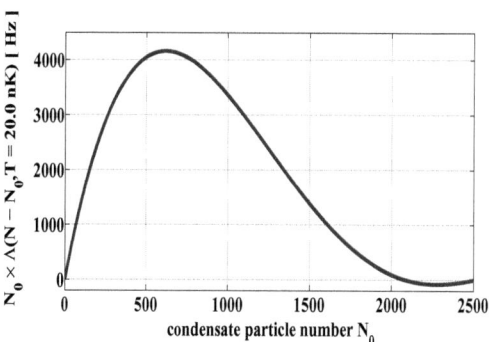

Figure 7.1: *Stimulated net single particle transition rate $N_0 \times \Lambda(N-N_0, T)$ as a function of $N_0$, for $N = 2500$ $^{87}$Rb atoms in a three-dimensional harmonic trap with frequencies $\omega_x = \omega_y = 2\pi \times 42.0$ Hz, $\omega_z = 2\pi \times 120.0$ Hz, with $\Lambda(N-N_0, T) = (\lambda_{\leadsto}^{+}(N-N_0, T) - \lambda_{\leadsto}^{-}(N-N_0, T))$. The final temperature of the gas is $T = 20.0$ nK. The critical temperature is $T_c = 36.47$ nK. Formation of an initially empty Bose-Einstein condensate corresponds to a path from left to the right in the figure until the reaching of equal particle balance $\Lambda(N-N_0, T) = 0$. Exact balance (at the intersection with the ordinate) is never reached completely (because the net feeding rate decreases to zero while approaching the intersection), but is approached arbitrarily close after few seconds.*

for the single particle feeding and loss rates.[1] Herein, we introduced the frequencies $\omega_k \equiv \epsilon_k/\hbar$. The $\delta$-function in Eq. (7.3),

$$\delta^{(\Gamma)}(\Delta\omega) = \frac{\sqrt{\pi}}{\Gamma} \exp\left[-\frac{(\Delta\omega)^2}{4\Gamma^2}\right], \tag{7.4}$$

originates from energy conservation and assumes a finite energy width $\sim \Gamma = \tau_{col}^{-1}$ according to the temporal decay of non-condensate phase correlations.

---

[1] Note that, Eq. (7.3) contains an infinite sum over all microscopic single particle exchange processes as depicted in the upper right diagrams in Fig. 5.1, hence relating the feeding rate in number representation to the spatial and to the thermal, coarse-grained average over all possible spatial configurations of the gas particles.

The weight function $\mathscr{F}_{\rightsquigarrow}^{\pm}(k,l,m,N-N_0,T)$ for single particle feeding processes is given by

$$\mathscr{F}_{\rightsquigarrow}^{+}(k,l,m,N-N_0,T) = f_k(N-N_0,T)f_l(N-N_0,T)[f_m(N-N_0,T)+1]|\zeta_{kl}^{m0}|^2 , \quad (7.5)$$

whereas the function $\mathscr{F}_{\rightsquigarrow}^{-}(k,l,m,N-N_0,T)$ for single particle losses turns into

$$\mathscr{F}_{\rightsquigarrow}^{-}(k,l,m,N-N_0,T) = [f_k(N-N_0,T)+1][f_l(N-N_0,T)+1]f_m(N-N_0,T)|\zeta_{m0}^{lk}|^2 . \quad (7.6)$$

In Eqs. (7.5, 7.6), $f_k(N-N_0,T)$ denotes the average occupation number of a non-condensate single particle mode $|\Psi_k\rangle$,

$$f_k(N-N_0,T) = \frac{1}{\exp[\beta(\epsilon_k - \mu_\perp(N-N_0,T))]-1} , \quad (7.7)$$

where $\mu_\perp(N-N_0,T)$ is implicitly defined by the normalization condition for the average non-condensate single particle occupations,

$$\sum_{k\neq 0} f_k(N-N_0,T) = (N-N_0) , \quad (7.8)$$

according to the fact that $N-N_0$ particles populate non-condensate modes (for a detailed discussion and derivation of Eqs. (7.7, 7.8), consult Section 7.2 and Appendix A.3). Since the particles are indistinguishable, each two body process with energy balance $\epsilon_k + \epsilon_l \simeq \epsilon_m + \epsilon_0$ is weighted by the corresponding occupation numbers $f_k(N-N_0,T), f_l(N-N_0,T), f_m(N-N_0)+1$ and $N_0+1$, and vice versa. The probability amplitudes $(\zeta_{m0}^{lk})^* = \zeta_{kl}^{0m}$ for single particle transitions occuring in the weight functions $\mathscr{F}_{\rightsquigarrow}^{\pm}(k,l,m,N-N_0,T)$ in Eqs. (7.5, 7.6) take into account the quantum mechanical

## 7.1. SINGLE PARTICLE (∼), PAIR (∼∼) AND SCATTERING (◯) RATES

wave nature of the particles. The average over the particles' waves functions is carried out in position space, the transition amplitudes being specified by overlap integrals over single particle states,

$$\zeta_{kl}^{m0} = \int_{\mathscr{C}} d\vec{r}\, \Psi_0^*(\vec{r})\Psi_m^*(\vec{r})\Psi_k(\vec{r})\Psi_l(\vec{r}) \,. \tag{7.9}$$

Note that the spatial average over all positions of the quantum matter waves in the gas as well as the coherent time evolution (on the finite coherence time $\tau_{col}$) of the waves according to the propagators in Eq. (7.1) is taken into account to calculate single particle transition rates.

Equations (7.9, 7.8, 7.7) are sufficient to quantify the weight functions $\mathscr{F}_{\sim}^{\pm}(k,l,m,N-N_0,T)$ in Eqs. (7.5, 7.6) and therewith the single particle feeding and loss rates $\lambda_{\sim}^{\pm}(N-N_0,T)$ in Eq. (7.3) numerically for all $k,l,m$, and for any state of $N-N_0$ particles: after explicit calculation of the overlap integrals $\zeta_{kl}^{m0}$ and $(\zeta_{kl}^{m0})^*$ for all $k,l,m$, $\mu_{\perp}(N-N_0,T)$ can be solved numerically by the normalization condition (7.8) for each $N-N_0$, leading to $\mathscr{F}_{\sim}^{\pm}(k,l,m,N-N_0,T)$ with the single particle occupations in Eq. (7.7). The single particle energies $\epsilon_k$ are defined by the diagonalization procedure in Section 4.3. Explicit expressions for all two body transition rates and energy shifts are given in Section 7.5 for a three-dimensional harmonic trap in the perturbative limit.

### 7.1.2 Pair feeding and loss rates

The calculation of the pair feeding and loss rates $\lambda_{\sim\sim}^{\pm}(N-N_0,T)$ in Eqs. (5.51, 5.52) is performed in the same spirit. However, it will be shown that – in contrast to single particle rates – the pair rates lead to a negligible contribution, if the condensate chemical potential $\mu_0$ lies energetically below the energies of single particle non-condensate states (which is the case for three-dimensional trap geometries in dilute atomic gases [17, 18, 95]).

Pair events describe the simultaneous exchange of two particles between condensate and non-condensate, quantified by the pair feeding and loss rates

$$\lambda^{\pm}_{\rawave}(N-N_0,T) = \frac{g^2}{8\hbar^2} \iint_{\mathscr{C}\times\mathscr{C}} d\vec{r}\, d\vec{r}^{\,\prime}\, \Psi_0(\vec{r})\Psi_0(\vec{r})\Psi_0^*(\vec{r}^{\,\prime})\Psi_0^*(\vec{r}^{\,\prime}) \int_{-\infty}^{\infty} d\tau\, e^{\pm 2i\omega_0\tau - \Gamma^2\tau^2}\, \mathscr{G}^{(\pm)}_{\rawave}(\vec{r},\vec{r}^{\,\prime},N-N_0,T,\tau)\,, \tag{7.10}$$

where we used that

$$\mathscr{G}^{(\pm)}_{\rawave}(\vec{r},\vec{r}^{\,\prime},N-N_0,T,\tau) = \left[\mathscr{G}^{(\mp)}_{\rawave}(\vec{r},\vec{r}^{\,\prime},N-N_0,T,-\tau)\right]^*\,. \tag{7.11}$$

Equations (A.11, A.12) of Appendix A.1 yield the explicit analytical expressions for pair feeding and loss rates,

$$\lambda^{\pm}_{\rawave}(N-N_0,T) = \frac{2\pi^3 \hbar^2 a^2}{m^2} \sum_{k,l\neq 0} \mathscr{F}^{\pm}_{\rawave}(k,l,N-N_0,T)\delta^{(\Gamma)}(\omega_k + \omega_l - 2\omega_0)\,, \tag{7.12}$$

where the weight function for pair feedings is given by

$$\mathscr{F}^{+}_{\rawave}(k,l,N-N_0,T) = f_k(N-N_0,T)f_l(N-N_0,T)|\zeta^{00}_{kl}|^2\,, \tag{7.13}$$

and, correspondingly, the weight function for pair losses turns into

$$\mathscr{F}^{-}_{\rawave}(k,l,N-N_0,T) = [f_k(N-N_0,T)+1][f_l(N-N_0,T)+1]|\zeta^{00}_{kl}|^2\,, \tag{7.14}$$

with probability amlitudes $\zeta^{00}_{kl} = \int d\vec{r}\,(\Psi_0^2(\vec{r}))^*\Psi_k(\vec{r})\Psi_l(\vec{r})$.

Looking at the energy balance of a pair event, $\Delta\epsilon_{\rawave} = \epsilon_k + \epsilon_l - 2\mu_0$, pair processes are obviously not energy conserving, meaning that $\Delta\epsilon_{\rawave} > \hbar\Gamma$, considering that the condensate chemical potential $\mu_0$ lies energetically below the energies of the single particle excited states [17, 18, 95]. Hence, the rates for pair events ($\rawave$) are negligible as compared to single particle events ($\leadsto$). This is in agreement

## 7.1. SINGLE PARTICLE (↝), PAIR (↜) AND SCATTERING (↻) RATES

with experimental observations [95], showing that dominant two body interaction processes are single particle processes (↝) in dilute quantum degenerate atomic gases.

### 7.1.3 Two body scattering rates

For the sake of completeness, the previous calculus of Sections 7.1.1, 7.1.2 is finally applied for calculating scattering rates between condensate and non-condensate particles in the Bose gas, which are formally defined by Eqs. (5.61, 5.62). Although scattering processes do not contribute to the quantum master equation in Eq. (5.68), it is physically interesting to calculate the order of magnitude of atomic scattering processes with $\Delta N_0 = \Delta N_\perp = 0$ in a Bose gas, because they contribute (additionally to the thermalization process) to decohering off-diagonal elements of the reduced condensate density matrix in Fock number representation (with a rate proportional to $N_0^2 \lambda_\circlearrowleft(N-N_0,T)$). The rate for scattering events reads:

$$\lambda_\circlearrowleft(N-N_0,T) = \frac{2g^2}{\hbar^2} \iint_{\mathscr{C}\times\mathscr{C}} d\vec{r}\, d\vec{r}'\, |\Psi_0(\vec{r})|^2 |\Psi_0(\vec{r}')|^2 \int_{-\infty}^{\infty} d\tau\, e^{-\Gamma^2\tau^2} \mathscr{G}_\circlearrowleft(\vec{r},\vec{r}',N-N_0,T,\tau)\,. \quad (7.15)$$

The decomposition of the correlation function $\mathscr{G}_\circlearrowleft(\vec{r},\vec{r}',N-N_0,T,\tau)$ for scattering processes, carried out in Eq. (A.13) of Appendix A.1, leads to the following explicit analytical expression for the scattering rate:

$$\lambda_\circlearrowleft(N-N_0,T) = \frac{32\pi^3 \hbar^2 a^2}{m^2} \sum_{k,l\neq 0} \mathscr{F}_\circlearrowleft(k,l,N-N_0,T) \delta^{(\Gamma)}(\omega_k - \omega_l)\,, \quad (7.16)$$

where

$$\mathscr{F}_\circlearrowleft(k,l,N-N_0,T) = f_k(N-N_0,T)[f_l(N-N_0,T)+1]|\zeta_{k0}^{l0}|^2\,, \quad (7.17)$$

given the probability amplitudes $\zeta_{k0}^{l0} = \int_\mathscr{C} d\vec{r}\, |\Psi_0(\vec{r})|^2 \Psi_l^*(\vec{r})\Psi_k(\vec{r})$. Since scattering

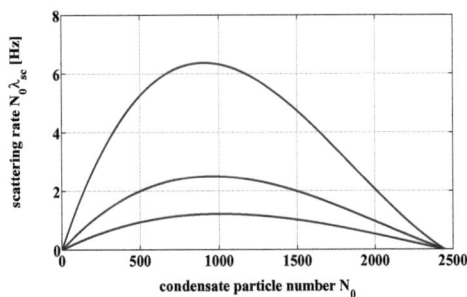

Figure 7.2: *Scattering rate $N_0 \lambda_\circlearrowleft (N - N_0, T)$ as a function of condensate particle number $N_0$, for $N = 2500$ $^{87}$Rb atoms in a three-dimensional harmonic trap with frequencies $\omega_x = \omega_y = 2\pi \times 42.0$ Hz, $\omega_z = 2\pi \times 120.0$ Hz. The corresponding critical temperature of the gas is $T_c = 36.47$ nK. The scattering rate is calculated for different temperatures $T = 20.0, 25.0$ and $30.0$ nK (from bottom to top). Scattering events erase off-diagonal elements of the reduced condensate density matrix in Fock number representation with a rate proportional to $N_0^2 \lambda_\circlearrowleft (N - N_0, T)$ additionally to the thermalization process, and therefore support the N-body Born-Markov ansatz (see Chapter 3).*

events do not change the particle number in the two subsystems condensate and non-condensate, the corresponding energy balances are $\Delta \epsilon_\circlearrowleft = \epsilon_k - \epsilon_l = 0$, as evident from the scattering rate in Eq. (7.16). The scattering rate is shown as a function of the condensate particle number in Fig. 7.2.

## 7.2 Depletion of the non-condensate

The function $\mu_\perp(N - N_0, T)$ in Eq. (7.8) occurs naturally in the derivation of average single particle occupations $f_k(N - N_0, T)$ of the non-condensate [10] in Eq. (7.7) (see Appendix A.3 for the derivation), and normalizes non-condensate single particle occupations to $(N - N_0)$ particles, for each condensate population of $N_0$ particles:

$$\sum_{k \neq 0} f_k(N - N_0, T) = \sum_{k \neq 0} \frac{1}{e^{\beta(\epsilon_k - \mu_\perp(N - N_0, T))} - 1} = (N - N_0) \, . \tag{7.18}$$

The dependence of non-condensate single particle occupations $f_k(N-N_0,T)$ on $(N-N_0)$, needed for the calculation of two body transition rates during Bose-Einstein condensation, is hence entirely determined by the normalization condition (7.18). Since each subspace of $N-N_0$ particles is a thermal mixture projected onto the Fock subspace of $N-N_0$ particles, $\mu_\perp(N-N_0,T)$ can also be interpreted as an ensemble of chemical potentials for the non-condensate [10] in dependence of $(N-N_0)$. Given analytically as

$$\mu_\perp(N-N_0) = -\beta^{-1}\frac{\partial \ln \mathcal{Z}_\perp(N-N_0)}{\partial(N-N_0)}, \qquad (7.19)$$

the function $\mu_\perp(N-N_0)$ is defined by the non-condensate partition function $\mathcal{Z}_\perp(N-N_0)$ in Eq. (5.15), and the final temperature $\beta = k_B T$ of the gas. Hence, $\mu_\perp(N-N_0)$ is proportional to the derivative of the Boltzmann entropy, and the Helmholtz free energy of the non-condensate part of the gas (compare chapter 1).

## 7.3 Detailed particle balance conditions

Emphasis is to be put on the balance conditions between loss and feeding rates for single particle processes in Eq. (7.3), and for pair processes in Eq. (7.12). As proven in Appendix A.2, the balance between single particle feeding and loss rates in Eq. (7.3) is

$$\lambda^+_\leadsto(N-N_0,T) = \exp[\beta\Delta\mu(N-N_0,T)]\lambda^-_\leadsto(N-N_0,T), \qquad (7.20)$$

valid for finite correlation times between condensate and non-condensate, $\hbar\Gamma\beta \ll 1$. In Eq. (7.20), $\Delta\mu(N-N_0) \equiv \mu_\perp(N-N_0,T) - \mu_0$ marks the difference between the eigenvalue of the Gross-Pitaevskii equation, $\mu_0$ in Eq. (4.4), and $\mu_\perp(N-N_0,T)$, which normalizes the thermal non-condensate single particle occupations, see Eqs. (7.8, 7.18).

Equation (7.20) explains the modulation of the particle balance between single particle feedings and losses between the condensate and the non-condensate modes in the gas by the difference $\Delta\mu_\perp(N-N_0,T)$. In particle number representation, the condensate mode is dynamically populated until energy balance $\mu_0 = \mu_\perp(N-N_0,T)$ between condensate and non-condensate is reached. As shown later (in Chapter 8), the steady state condition after condensate formation implies that $\mu_0 = \mu_\perp(N-\langle N_0\rangle,T)$ close to the maximum $\langle N_0\rangle$ of the steady state condensate number distribution, i.e. energetic equilibrium on average is established in the final equilibrium steady state: The equality of $\mu_0$ and $\mu_\perp$ means that the net average energy flow between condensate and non-condensate is zero [10] – in agreement with thermodynamics. At detailed balance, $\lambda^+_{\sim\sim}(N-\langle N_0\rangle,T) = \lambda^-_{\sim\sim}(N-\langle N_0\rangle,T)$, the particle exchange between condensate and non-condensate is stationary on average.

Similarly to single particle processes, also pair losses and feedings are not independent, but obey the balance condition

$$\lambda^+_{\sim\sim}(N-N_0,T) = \exp[2\beta\Delta\mu(N-N_0)]\lambda^-_{\sim\sim}(N-N_0,T) , \qquad (7.21)$$

showing that also the net exchange of particle pairs between condensate and non-condensate is zero, if $\mu_0 = \mu_\perp(N-N_0,T)$. In turn, also the reaching of a equal particle balance implies energetic equilibrium on average. At stationary particle flow, the energy flow is thus also conserved, if pair processes became resonant (for $\mu_0 > \epsilon_k$).

## 7.4 Single particle, pair and scattering energy shifts

Single particle, pair and scattering processes only contribute to the Lindblad dynamics if they are energy conserving. A two body collision event which does not microscopically conserve the energy, however, can still occur as a virtual process [20] (in conjunction with its conjugate process), if the time scale of this process is sufficiently fast (i.e., if it occurs on a time scale $\tau \ll \Delta\epsilon/\hbar\Gamma^2$). This is similar to

## 7.4. SINGLE PARTICLE, PAIR AND SCATTERING ENERGY SHIFTS

the Lamb shift in quantum optics [86], where fluctuations of the vacuum photon field induce a splitting of the angular momentum degenerate eigenenergies in the hydrogen atom.

According to the derivation of the Lindblad master equation in Chapter 5, the energy shift term describing the net effect of virtual processes is composed of the different principal parts of the complex valued transition rates:

$$\hat{\Delta}(N-N_0,T) = \hbar\Delta^{(lin)}(N-N_0,T)\hat{N}_0 + \hbar\Delta^{(nlin)}(N-N_0,T)\hat{N}_0^2 , \qquad (7.22)$$

where the quantities $\Delta^{(lin)}(N-N_0,T)$ and $\Delta^{(nlin)}(N-N_0,T)$ denote energy shifts,[2] which occur linearly and nonlinearly in the condensate number operator $\hat{N}_0$. Energy shifts $\Delta^{(lin)}(N-N_0,T)$ renormalize the single particle term $\vec{p}^2/2m + V_{ext}(\vec{r}) \to \vec{p}^2/2m + V_{ext}(\vec{r}) + \hbar\Delta^{(lin)}$, whereas nonlinear shifts renormalize the interaction energy $g|\Psi_0(\vec{r})|^2 \to g|\Psi_0(\vec{r})|^2 + \hbar\Delta^{(nlin)}$.

This renormalization can be used to add second order backreactions (in $g$) induced by virtual processes of the non-condensate field to the finite temperature Gross-Pitaevskii equation [46]:

$$\left\{ \left[ -\frac{\hbar^2\vec{\nabla}^2}{2m} + \hbar\Delta^{(lin)} + V_{ext}(\vec{r}) \right] + 2gn_{NC}(\vec{r}) + gn_C(\vec{r}) + N_0\hbar\Delta^{(nlin)} \right\} \Psi_0(\vec{r}) = \mu_0\Psi_0(\vec{r}) , \qquad (7.23)$$

where $n_{NC}(\vec{r})$ is the average non-condensate density, and $n_C(\vec{r})$ the average condensate density.

The renormalization of the eigenvalue $\mu_0$ due to $\hbar\Delta^{(lin)}$ and $\hbar\Delta^{(nlin)}$ is calculated numerically: the order of magnitude is about 1% of the single particle ground state energy $\hbar\omega$ for dilute and sufficiently small atomic gases, as displayed in Figs. 7.4, 7.5. Energy shifts are thus small in dilute atomic gases. For completeness, all components of the energy shifts in Eq. (7.23) are calculated explicitly,

---
[2] Specified in units of $s^{-1}$

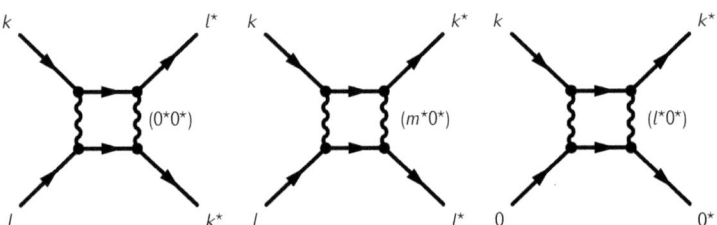

Figure 7.3: *Virtual single particle, pair and scattering processes create a virtual state. Resulting positive energy shifts, $\Delta^+_{\rightsquigarrow}$, $\Delta^+_{\rightsquigarrow\rightsquigarrow}$ and $\Delta^+_{\circlearrowright}$ in Eqs. (7.26, 7.27, 7.28) (from left to right), are represented by the connection of two conjugate single particle ($\rightsquigarrow$), pair ($\rightsquigarrow\rightsquigarrow$) and scattering ($\circlearrowright$) diagrams. Only positive energy shifts are displayed, negative shifts are obtained by the conjugate diagrams.*

originating from virtual processes associated with the different nature of single particle ($\rightsquigarrow$), pair ($\rightsquigarrow\rightsquigarrow$) and scattering processes ($\circlearrowright$).

First, we note that the total energy shift $\Delta^{(\mathrm{lin})}(N-N_0,T)$, which is linear in the condensate number operator $\hat{N}_0$, is composed of

$$\Delta^{(\mathrm{lin})}(N-N_0,T) = \Delta^+_{\rightsquigarrow}(N-N_0,T) + \Delta^-_{\rightsquigarrow}(N-N_0,T) + 4\Delta^+_{\rightsquigarrow\rightsquigarrow}(N-N_0,T) + \Delta^{\circlearrowright}(N-N_0,T) \,. \quad (7.24)$$

Besides, the shift $\Delta^{(\mathrm{nlin})}(N-N_0,T)$ which occurs nonlinearly in $\hat{N}_0$ is given by:

$$\Delta^{(\mathrm{nlin})}(N-N_0,T) = \Delta^+_{\rightsquigarrow\rightsquigarrow}(N-N_0,T) + \Delta^-_{\rightsquigarrow\rightsquigarrow}(N-N_0,T) + \Delta_{\circlearrowright}(N-N_0,T) \,. \quad (7.25)$$

Thus, each of the various processes (single particle, pair and scattering) induces positive (+) and negative (-) energy shifts. The two body processes leading to the energy shifts in Eqs. (7.24, 7.25) are depicted in Fig. 7.3, and are explicitly given by the imaginary parts of the rates in Eqs. (7.3, 7.12, 7.16). Single particle ($\rightsquigarrow$) energy

## 7.4. SINGLE PARTICLE, PAIR AND SCATTERING ENERGY SHIFTS

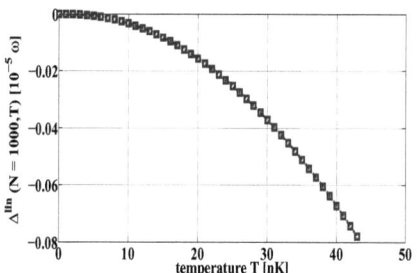

Figure 7.4: *Energy shift* $\Delta^{(\text{lin})} = \Delta_+^{\leftrightarrow}(N-N_0, T) + \Delta_-^{\leftrightarrow}(N-N_0, T) + 4\Delta_+^{\leadsto}(N-N_0, T) + \Delta^{\circlearrowleft}(N-N_0, T)$ *per particle in Eq. (7.24), in units of the trapping frequency* $\omega = 2\pi \times 600.0$ *Hz is displayed as a function of temperature T for* $N = 1000$ $^{87}$*Rb atoms in the trap.*

shifts hence turn into

$$\Delta_{\leftrightarrow}^{\pm}(N-N_0, T) = \frac{\pm 8\pi^2 a^2 \hbar^2}{m^2} \sum_{k,l,m \neq 0} \mathscr{F}_{\leftrightarrow}^{\pm}(k,l,m,N-N_0,T) \, \text{PV}\left\{\frac{1}{(\omega_l + \omega_m - \omega_k - \omega_0)}\right\}, \quad (7.26)$$

where PV$\{X\}$ labels the principle part of $X$ [58], whereas energy shifts for pair events ($\leadsto$) are given by

$$\Delta_{\leadsto}^{\pm}(N-N_0, T) = \frac{\pm \pi^2 a^2 \hbar^2}{m^2} \sum_{k,l \neq 0} \mathscr{F}_{\leadsto}^{\pm}(k,l,N-N_0,T) \, \text{PV}\left\{\frac{1}{(\omega_k + \omega_l - 2\omega_0)}\right\}. \quad (7.27)$$

Finally, the quantity

$$\Delta_{\circlearrowleft}(N-N_0, T) = \frac{-32\pi^2 a^2 \hbar^2}{m^2} \sum_{k,l \neq 0} \mathscr{F}_{\circlearrowleft}(k,l,N-N_0,T) \, \text{PV}\left\{\frac{1}{(\omega_k - \omega_l)}\right\} \quad (7.28)$$

denotes the energy shift induced by scattering processes ($\circlearrowleft$).

162    Chapter 7. TRANSITON RATES FOR BOSE-EINSTEIN CONDENSATION

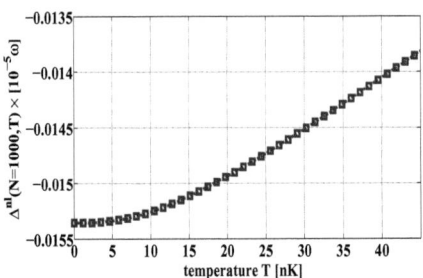

Figure 7.5: *Nonlinear energy shift per particle in units of the trapping frequency $\omega = 2\pi \times 600.0$ Hz as a function of temperature T, for $N = 1000$ $^{87}$Rb atoms in the trap. The total energy shift $\Delta^{(\text{nlin})}$ modifies the nonlinear term of the condensate Hamiltonian by a relative (to $g|\Psi_0|^2 \sim 10^{-5} - 10^{-3}\hbar\omega$) amount up to 1.0%.*

## 7.5 Transition rates and energy shifts in the perturbative limit

The small parameter $\xi = a\varrho^{1/3} \ll 1$ of our theory[3] can be identified in the transition rates in Eqs. (7.3, 7.12, 7.16), as well as in the energy shifts in Eqs. (7.26, 7.27, 7.28). The previously presented transition rates and energy shifts are defined via the Gross-Pitaevskii condensate wave function $|\Psi_0\rangle$ in Eq. (7.3), being a function of the product $gN$. The Gross-Pitaevskii state $|\Psi_0\rangle$ and therewith the single particle basis states $\{|\Psi_k\rangle, k \in \mathbb{N}\}$ being orthogonal to $|\Psi_0\rangle$ are thus functions of the parameter $a\varrho = \xi\varrho^{2/3}$, provided that the atomic density $\varrho$ is replaced by the peak density $N|\Psi_0(0)|^2$ at the center of the trap at sufficiently low temperatures.

Since the single particle energies $\epsilon_k$ in Eq. (4.28) and the corresponding number occupations $f_k(N - N_0, T)$ of non-condensate single particle states in Eq. (7.7) are defined in terms of the basis states $\{|\Psi_k\rangle, k \in \mathbb{N}\}$ (which can be expanded in the small parameter $\xi$), all of the two body transition rates (and the corresponding energy shifts) are composed of:

---
[3]Realized in many state-of-the-art experiments (see Section 1.4).

## 7.5. TRANSITION RATES AND ENERGY SHIFTS IN THE PERTURBATIVE LIMIT

$$\mathscr{X}(N-N_0, T) = \mathscr{X}(a^2, N-N_0, T) + \mathcal{O}(a^3) , \qquad (7.29)$$

where the leading order contribution $\mathscr{X}(a^2, N-N_0, T)$ is proportional to $a^2$. The $k^{\text{th}}$ correction to the transition rates and energy shifts scales as $(a\varrho^{1/3})^k \to 0^+$ relatively to the leading order contribution $\mathscr{X}(a^2, N_0, T)$.

Hence, the diluteness of a Bose-Einstein condensate, formally reflected by the dilute gas condition $a\varrho^{1/3} \ll 1$, enters as a perturbation theory for single particle wave functions into the master equation formalism: In dilute atomic gases, interactions are sufficiently weak to replace the wave functions of the interacting particles by the ones of an ideal gas in a master equation governing the dynamics of two body collisions (proportional to $a^2$). This justifies from first principles the use of single particle states and single particle energies of a non-interacting gas in the QBE (compare the transition rates in Eq. (2.17)).

Not least for numerical simplicity, we therefore restrict quantitative predictions to the formal limiting case $a\varrho^{1/3} \to 0^+$ (with $a \neq 0$ of $\varrho = $ const.) of weak interactions, taking into account only the leading order of transition rates, which are still proportional to $a^2$. Note that the rates obtained in the perturbative limit still contain an infinite series over all two body interaction processes, being perturbative from the point of view that the disturbance of the single particle wave functions is omitted.

### 7.5.1 Leading order of transition rates

The leading order contributions of the two body transition rates $\lambda_j^{\mp}(N-N_0, T)$ are calculated for $j = \rightsquigarrow, \leftrightsquigarrow, \circlearrowright$ for a three-dimensional harmonic trap with spatial extentions $L_\eta = \sqrt{\hbar/m\omega_\eta}$, where $\omega_{x,y,z} = 2\pi \times \nu_{x,y,z}$ ($\eta = x, y, z$) are the trap frequencies.

According to Section 7.5, the leading order contribution of transition rates and energy shifts are quantified by the Schrödinger equation:

$$\sum_{\eta=x,y,z}\left(\frac{\hat{p}_\eta^2}{2m}+\frac{1}{2}\omega_\eta\hat{\eta}^2\right)|\chi_{\vec{k}}\rangle=\epsilon_{\vec{k}}|\chi_{\vec{k}}\rangle. \tag{7.30}$$

Equation (7.30) can be solved analytically exactly, leading to single particle eigenstates of the non-interacting system:

$$\langle\vec{r}|\chi_{\vec{k}}\rangle=\mathcal{N}\prod_{\eta=x,y,z}\sqrt{\frac{1}{L_\eta}}\exp\left[-\frac{\eta^2}{2L_\eta^2}\right]H_{k_\eta}(L_\eta\eta), \tag{7.31}$$

with $H_{k_\eta}(L_\eta\eta)$, the Hermite polynomials [40], and $\mathcal{N}=(\pi^{3/4}\sqrt{2^{k_x+k_y+k_z}}k_x!k_y!k_z!)^{-1}$, a normalization constant. The corresponding quantized single particle eigenenergies are

$$\epsilon_{\vec{k}}=\left(k_x+\frac{1}{2}\right)\hbar\omega_x+\left(k_y+\frac{1}{2}\right)\hbar\omega_y+\left(k_z+\frac{1}{2}\right)\hbar\omega_z, \tag{7.32}$$

with $\vec{k}=(k_x,k_y,k_z)^T$. Since Eqs. (7.7, 7.8) fully specify single particle occupations $f_{\vec{k}}(N-N_0,T)$ with respect to the single particle states and single particle energies in Eqs. (7.31, 7.32), the leading order contributions of the transitions rates $\lambda_{\leadsto}^\pm(N-N_0,T)$, $\lambda_{\leadsto\leadsto}^\pm(N-N_0,T)$ and $\lambda_{\circleddash}^\pm(N-N_0,T)$, as well as of the energy shifts $\Delta_{\leadsto}^\pm(N-N_0,T)$, $\Delta_{\leadsto\leadsto}^\pm(N-N_0,T)$ and $\Delta_{\circleddash}^\pm(N-N_0,T)$ can be calculated analytically. The leading order feeding and loss rate[4] associated to single particle processes ($\leadsto$) turns, after calculation [96] of the overlap integrals in Eq. (7.3), into:

$$\lambda_{\leadsto}^\pm(N-N_0,T)=\frac{2ma^2\bar{\omega}}{\hbar\pi^9}\sum_{\vec{k},\vec{l},\vec{m}\neq 0}\mathscr{F}_{\leadsto}^\pm(\vec{k},\vec{l},\vec{m},N-N_0,T)\,\delta^{(\Gamma)}\left((\vec{k}+\vec{l}-\vec{m})\cdot\vec{\omega}\right), \tag{7.33}$$

with $\vec{\omega}=(\omega_x,\omega_y,\omega_z)$, and $\bar{\omega}=(\omega_x\omega_y\omega_z)^{1/3}$. The weight functions $\mathscr{F}_{\leadsto}^\pm(\vec{k},\vec{l},\vec{m},N-N_0,T)$ in Eq. (7.33) are given by

---
[4]We do not introduce an extra label for brevity.

## 7.5. TRANSITION RATES AND ENERGY SHIFTS IN THE PERTURBATIVE LIMIT

$$\mathscr{F}^+_{\leadsto}(\vec{k},\vec{l},\vec{m},N-N_0,T) = f_{\vec{k}}(N-N_0,T)f_{\vec{l}}(N-N_0,T)[f_{\vec{m}}(N-N_0,T)+1]\Sigma_{\leadsto}(\vec{k},\vec{l},\vec{m}),\quad (7.34)$$

for single particle feedings. The function $\Sigma_{\leadsto}(\vec{k},\vec{l},\vec{m}) = 0$, if $(k_\eta + l_\eta + m_\eta)$ is odd ($\eta = x,y,z$), and turns into

$$\Sigma_{\leadsto}(\vec{k},\vec{l},\vec{m}) = \prod_{\eta=x,y,z}\left[\frac{\Gamma\left(k_\eta+l_\eta+\tfrac{1}{2}\right)\Gamma\left(m_\eta+k_\eta+\tfrac{1}{2}\right)\Gamma\left(l_\eta+m_\eta+\tfrac{1}{2}\right)}{\sqrt{k_\eta!l_\eta!m_\eta!}}\right]^2 \quad (7.35)$$

otherwise, because of the alternating parity of the harmonic oscillator states. In Eq. (7.35), $\Gamma(x)$ is the Euler Gamma function [40].

The weight function $\mathscr{F}^-_{\leadsto}(\vec{k},\vec{l},\vec{m},N-N_0,T)$ for single particle loss processes turns into

$$\mathscr{F}^-_{\leadsto}(\vec{k},\vec{l},\vec{m},N-N_0,T) = [f_{\vec{k}}(N-N_0,T)+1][f_{\vec{l}}(N-N_0,T)+1]f_{\vec{m}}(N-N_0,T)\Sigma_{\leadsto}(\vec{k},\vec{l},\vec{m}).\quad (7.36)$$

The dependence of the single particle loss and feeding rates in Eq. (7.33) on $N_0$ is displayed in Fig. 7.6. The difference between losses and feedings disappears for a certain value of $N_0$ (marking the detailed balance particle flow, see Eq. (7.20)), which is determined by thermodynamical constraints such as temperature, volume of the trap and the total particle number (see Chapter 8).

We find the typical magnitude of the single particle feeding and loss rates to be $0.1\ldots100$ Hz, see Fig. 7.6: For low final temperatures ($T = 20.0$ nK), single particle feedings are fast ($\sim 60$ Hz) for weakly occupied condensates, the gas being far from equilibrium. In consequence, the net feeding rate $\lambda^+_{\leadsto}(N-N_0,T) - \lambda^-_{\leadsto}(N-N_0,T)$ remains large and positive up to $N_0 \sim 2200$, and smaller below (meaning that the bose-condensed state is dynamically stable). At high final temperatures ($T = 35.0$ nK) feeding rates are larger than loss rates until $N_0 \sim 150$, and smaller below.

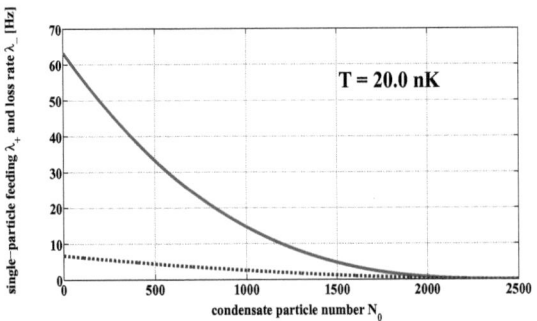

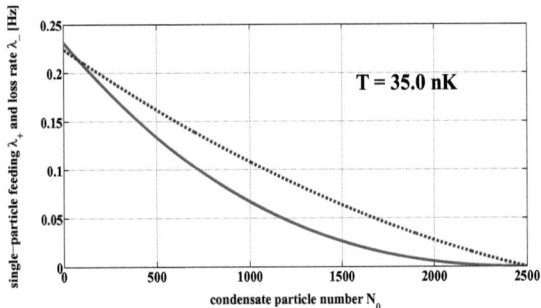

Figure 7.6: *Single particle feeding $\lambda_{\leadsto}^{+}(N-N_0, T)$ (red solid line) and loss rates $\lambda_{\leadsto}^{-}(N-N_0, T)$ (blue dashed line) of Eq. (7.33) are presented as a function of the condensate particle number $N_0$, for two different temperatures, $T = 20.0$ and $35.0$ nK, for a gas of $N = 2500$ $^{87}$Rb atoms in a three-dimensional harmonic trap, with frequencies $\omega_x = \omega_y = 2\pi \times 42.0$ Hz, $\omega_z = 2\pi \times 120.0$ Hz. The critical temperature of the gas is $T_c = 36.47$ nK. For $T = 20.0$ nK, the particle flow to the condensate is fast, highlighting the high non-equilibrium situation, whereas it is smaller by an order of magnitude for $T = 35.0$ nK. There is only one intersection of the feeding and loss rate (unique steady state), at $N_0^{is} = 2200$ for $T = 20.0$ nK, and at $N_0^{is} = 150$ for $T = 35.0$ nK, which marks the number $N_0$ around which the two subsystems exhibit detailed balance particle flow. Since $\lambda_{\leadsto}^{+}(N-N_0, T) - \lambda_{\leadsto}^{-}(N-N_0, T) > 0$ for $N_0 < N_0^{is}$ and $\lambda_{\leadsto}^{+}(N-N_0, T) - \lambda_{\leadsto}^{-}(N-N_0, T) < 0$ for $N_0 < N_0^{is}$, the steady state of the Bose-Einstein condensate is dynamically stable. Indeed, for decreasing (different) temperatures (not shown), the intersection point travels from $N_0^{is} = 0$ to $N_0^{is} = N$.*

## 7.5. TRANSITION RATES AND ENERGY SHIFTS IN THE PERTURBATIVE LIMIT

In both cases, detailed balance particle flow is obviously reached at the (only) intersection of $\lambda^+_{\sim\sim}(N-N_0,T)$ and $\lambda^-_{\sim\sim}(N-N_0,T)$, which hints already at this stage the approach to a unique and stable stationary state. The steady state is derived and analyzed in more detail in chapter 8.

The unperturbed pair feeding and loss rates turn into

$$\lambda^\pm_{\sim\sim}(N-N_0,T) = \frac{ma^2\bar{\omega}}{\hbar\pi^3}\sum_{\vec{k},\vec{l}\neq 0}\mathscr{F}^\pm_{\sim\sim}(\vec{k},\vec{l},N-N_0,T)\,\delta^{(\Gamma)}\left((\vec{k}+\vec{l})\cdot\vec{\omega}\right)\,, \qquad (7.37)$$

with

$$\mathscr{F}^+_{\sim\sim}(\vec{k},\vec{l},N-N_0,T) = \left[f_{\vec{k}}(N-N_0,T)+1\right]\left[f_{\vec{l}}(N-N_0,T)+1\right]\Sigma_{\sim\sim}(\vec{k},\vec{l})\,, \qquad (7.38)$$

and similarly for pair losses:

$$\mathscr{F}^-_{\sim\sim}(\vec{k},\vec{l},N-N_0,T) = f_{\vec{k}}(N-N_0,T)f_{\vec{l}}(N-N_0,T)\Sigma_{\sim\sim}(\vec{k},\vec{l})\,. \qquad (7.39)$$

In Eqs. (7.38, 7.39), the function $\Sigma_{\sim\sim}(\vec{k},\vec{l})$ is given by

$$\Sigma_{\sim\sim}(\vec{k},\vec{l}) = \prod_{\zeta=x,y,z}\left[\frac{\Gamma\left(\frac{k_\eta+l_\eta+1}{2}\right)}{\sqrt{k_\eta!l_\eta!}}\right]^2\,, \qquad (7.40)$$

for even $(k_\eta+l_\eta)$ $(\eta=x,y,z)$, and it is zero otherwise. Once again, Eq. (7.37) verifies that pair processes are off-resonant for sufficiently large trapping frequencies because of energy conservation, and do therefore not contribute to the dynamics of the Bose gas, see Eq. (7.37).

The leading order contribution for scattering rates finally turns into:

$$\lambda_\circlearrowleft(N-N_0,T) = \frac{2ma^2\bar{\omega}}{\hbar\pi^3}\sum_{\vec{k},\vec{l}\neq 0}\mathscr{F}_\circlearrowleft(\vec{k},\vec{l},N-N_0,T)\,\delta^{(\Gamma)}\left((\vec{k}+\vec{l})\cdot\vec{\omega}\right)\,, \qquad (7.41)$$

with a weight function $\mathscr{F}_\circlearrowleft(\vec{k},\vec{l},N-N_0,T)$ given by

$$\mathscr{F}_\circlearrowleft(\vec{k},\vec{l},N-N_0,T) = [f_{\vec{k}}(N-N_0,T)+1]f_{\vec{l}}(N-N_0,T) \prod_{\zeta=x,y,z} \left[\frac{\Gamma\left(\frac{k_\eta+l_\eta+1}{2}\right)}{\sqrt{k_\eta!l_\eta!}}\right]^2 . \quad (7.42)$$

The scattering rate $N_0\lambda_\circlearrowleft(N_0,T)$ is presented in Fig. 7.2 as a function of the condensate particle number $N_0$ with the same parameters as in Fig. 7.6, and for different temperatures $T = 20.0, 25.0, 30.0$ nK.

For numerical calculations throughout the thesis, these perturbative two body transition rates for single particle exchanges in Eq. (7.33) are employed from now on, accurately resembling the condensate formation times of state-of-the-art experiments with $\xi \sim 10^{-1}-10^{-2} \ll 1$ (see chapter 6).

### 7.5.2 Leading order energy shifts

As discussed in Section 7.4, energy non-conserving processes create intermediate states of short life time. The net effect comprises the energy shifts in Eqs. (7.24, 7.25), which renormalize the single particle ground state energy, see Eq. (7.23). Energy shift terms $\Delta^{(\text{lin})}$ and $\Delta^{(\text{nlin})}$ are quantified here for a three-dimensional harmonic trap in the perturbative limit $a\varrho^{1/3} \to 0^+$.

First, remember that unperturbed energetic shifts linear in $N_0$ are expressed by

$$\Delta^{(\text{lin})}(N-N_0,T) = \Delta^+_{\rightsquigarrow}(N-N_0,T)+\Delta^-_{\rightsquigarrow}(N-N_0,T)+4\Delta^+_{\rightsquigarrow}(N-N_0,T)+\Delta_\circlearrowleft(N-N_0,T) , \quad (7.43)$$

and the shift occuring nonlinearly in $N_0$ consists of

$$\Delta^{(\text{nlin})}(N-N_0,T) = \Delta^+_{\rightsquigarrow}(N-N_0,T)+\Delta^-_{\rightsquigarrow}(N-N_0,T)+\Delta_\circlearrowleft(N-N_0,T) . \quad (7.44)$$

## 7.6. GENERALIZED EINSTEIN DE BROGLIE CONDITION

Both energy shifts are thus simultaneously defined in terms of the different imaginary counterparts of the unperturbed rates in Eqs. (7.33, 7.37, 7.41). Starting with unperturbed single particle energy shifts ($\leadsto$), they turn into

$$\Delta_{\leadsto}^{\pm}(N-N_0, T) = \pm\frac{2ma^2\overline{\omega}^3}{\hbar\pi^{10}} \sum_{\vec{k},\vec{l},\vec{m}\neq 0} \mathscr{F}_{\leadsto}^{\pm}(\vec{k},\vec{l},\vec{m}, N-N_0, T) \, \text{PV}\left\{\frac{1}{(\vec{l}+\vec{m}-\vec{k})\cdot\vec{\omega}}\right\} . \quad (7.45)$$

The unperturbed energy shifts for pair events ($\leftrightsquigarrow$) are given by

$$\Delta_{\leftrightsquigarrow}^{\pm}(N-N_0, T) = \frac{\pm ma^2\overline{\omega}^3}{\hbar\pi^{10}} \sum_{\vec{k},\vec{l}\neq 0} \mathscr{F}_{\leftrightsquigarrow}^{\pm}(\vec{k},\vec{l}, N-N_0, T) \, \text{PV}\left\{\frac{1}{(\vec{k}+\vec{l})\cdot\vec{\omega}}\right\} , \quad (7.46)$$

and finally the expression

$$\Delta_{\circlearrowleft}(N-N_0, T) = \frac{-2ma^2\overline{\omega}^3}{\hbar\pi^4} \sum_{\vec{k},\vec{l}\neq 0} \mathscr{F}_{\circlearrowleft}(\vec{k},\vec{l}, N-N_0, T) \, \text{PV}\left\{\frac{1}{(\vec{k}-\vec{l})\cdot\vec{\omega}}\right\} \quad (7.47)$$

determines unperturbed energy shifts induced by scattering processes ($\circlearrowleft$). The numerical values of the energy shifts in Eqs. (7.45, 7.46, 7.47) are shown in Figs. 7.4, 7.5 of Section 7.4.

## 7.6 Generalized Einstein de Broglie condition

The density of states of an ideal gas ($g \to 0^+$) in the semiclassical limit, where $k_B T \gg \hbar\eta_0$, and $N \to \infty$, is given [15] by $g(\eta) = Vm^{3/2}/2^{1/2}\pi^2\hbar^3\eta^{1/2}$ for uniform gases, and by $g(\eta) = \eta^{d-1}/(d-1)!\prod_{j=x,y,z}\hbar\omega_j$ for $d$-dimensional harmonic traps with frequencies $\omega_j = 2\pi \times \nu_j$ ($j = x, y, z$). Integrating the number of non-condensate particles over the density of states $g(\eta)$, the closed, implicit equation for the non-condensate chemical potential $\mu_\perp(N-N_0, T)$ in Eq. (7.8) turns into

| $\gamma$ | $\Gamma(\gamma)$ | $\zeta(\gamma)$ |
|---|---|---|
| 1.0 | 1.0 | $\infty$ |
| 1.5 | $\sqrt{\pi}/2 \approx 0.886$ | 2.612 |
| 2.0 | 1.0 | $\pi^2/6 \approx 1.645$ |
| 2.5 | $3\sqrt{\pi}/4 \approx 1.329$ | 1.341 |
| 3.0 | 2.0 | 1.202 |
| 3.5 | $15\sqrt{\pi}/8 \approx 3.323$ | 1.127 |
| 4.0 | 6.0 | $\pi^4/90 \approx 1.082$ |

Table 7.1: *Euler Gamma function $\Gamma(\gamma)$, and the Riemann Zeta function $\zeta(\gamma)$, for selected values of $\gamma$.*

$$m_\gamma[e^{\beta\mu_\perp(N-N_0)}] = \frac{(N-N_0)}{C_\gamma \Gamma(\gamma)(k_B T)^\gamma}, \qquad (7.48)$$

with $\Gamma(\gamma)$, the Euler Gamma function. The parameter $\gamma = 3/2$ for uniform gases, and $\gamma = 3$ for a harmonic trap. $m_\gamma[z] = \sum_{n=1}^{\infty} z^n/n^\gamma$ is the Bose function [15]. At the Bose-Einstein phase transition, $z = 1$, and $m_\gamma[z \to 1] = \zeta(\gamma)$ reduces to the Zeta function. Some values for $\Gamma(\gamma)$ and $\zeta(\gamma)$ are summarized in Table 7.1. $C_\gamma$ in Eq. (7.48) corresponds to a geometry dependent constant, which is analytically given by $C_{3/2} = Vm^{3/2}2^{-1/2}\pi^{-2}\hbar^{-3}$ for the uniform case; for a $d$-dimensional harmonic trap, it follows that $C_d = 1/(d-1)! \prod_{i=1,...,d} \hbar\omega_i$ [15].

The approximate relation (which becomes exact for large particle numbers) in Eq. (7.48), together with the detailed balance condition in Eq. (7.20), bears an important physical interpretation.

At detailed balance (which is reached on average in the stationary state, see Section 8.1) the chemical potential of the non-condensate $\mu_\perp$ in Eq. (7.48) reaches the single particle ground state energy $\eta_0$. In the perturbative limit of vanishing interactions (see Section 7.5), the single particle ground state energy can be set to zero, because the dynamics and the statistics of the Bose gas is invariant under the transformation $\hat{\mathcal{H}} \to \hat{\mathcal{H}} - \eta_0 \hat{N}$ of the total Hamiltonian, where $\eta_0$ is the single particle ground state energy (the easiest check of this statement is to recognize that $\hat{\sigma}^{(N)}(t)$ commutes with $\hat{N}$ at any time $t$). For uniform gases ($\gamma = 3/2$), Bose-Einstein hence

## 7.6. GENERALIZED EINSTEIN DE BROGLIE CONDITION

sets in, if $(N-N_0)/C_{3/2}\Gamma(3/2)(k_B T)^{3/2}$ in Eq. (7.48) is larger than $\zeta(3/2)$ ($\mu_\perp > 0$), defining the condition $(N - N_0)/V\lambda^3(T) > \zeta(3/2)$ for Bose-Einstein condensation, stationary particle flow being reached if, and only if $(N - N_0)/V\lambda^3(T) = \zeta(3/2)$ ($\mu_\perp = 0$).

Using the above explicit expression for $C_{3/2}$, and identifying $\varrho_\perp = (N-N_0)/V$, Bose-Einstein condensation consequently manifests itself as a mechanism for the Bose gas to reduce its *non-condensate* density, until the ratio of thermal de Broglie volume and mean particle distance in the *non-condensate* equals $\zeta(3/2)$. For arbitrary temperatures (below $T_c$), we thus recover a condition for condensate formation (stationarity) similar to the Einstein de Broglie relation (1.3), but for the non-condensate part of the gas:

$$\varrho_\perp \lambda^3(T) \geq \zeta(3/2) \ . \tag{7.49}$$

On thermodynamic grounds, Eq. (7.49) implies that the entropy of the non-condensate gas is maximized, and that the free energy is minimized during condensate formation, see Eq. (1.2, 7.19).

From the wave mechanics point of view, Eq. (7.49) can be interpreted as Bose-Einstein condensation to occur as long as the non-condensate gas particles remain spatially coherent. Since the particle number and the trap volume during condensate formation are conserved, we see that, at equilibrium, the condensate fraction is that part of the gas particles exhibiting an average atomic density allowing to find more than one particle in the coherence volume $\lambda^3(T)$,

$$\varrho_0 \lambda^3(T) = \varrho\lambda^3(T) - \zeta(3/2) \ , \tag{7.50}$$

whereas the non-condensate gas particles are spaced on the order of $\lambda^3(T)$ on average.[5] Note, that Eqs. (7.49, 7.50) reduce to the usual Einstein de Broglie condition

---
[5] For a three-dimensional harmonic trap, the situation is qualitatively the same: The condensate thus forms until the right hand side of Eq. (7.48) equals $\zeta(3)$ at final equilib-

in Eq. (1.3) at the phase transition temperature, where $\varrho_\perp = \varrho$.

The dynamics of the Bose-Einstein phase transition after the quench of the non-condensate density, $\varrho_\perp \lambda^3(T) > \zeta(3/2)$ (or $(N-N_0)\hbar^3 \omega_x \omega_y \omega_0/(k_B T)^3 > \zeta(3)$ for a three-dimensional harmonic trap, respectively), was shown in Chapter 6, monitoring the behavior of the condensate and the non-condensate number distributions during condensate formation.

---

rium, i.e. for $\mu_\perp = 0$ (and is larger before the reaching of the detailed balance condition). However, it cannot be directly related to the ratio of de Broglie volume and average particle spacing, as the non-condensate density is not homogeneous because of the external trapping confinement. Also here is the phase transition independent on the choice of $\eta_0$, which can hence be set to zero.

Chapter 8

# Equilibrium properties of a dilute Bose-Einstein condensate

Analyzing the steady state solution of the master equation analytically proofs the existence of a unique steady state for dilute, weakly interacting Bose-Einstein condensates of finite particle number under the Markov dynamics assumption. In the limiting case of weak interactions, the latter is given by a Gibbs-Boltzmann thermal state of $N$ indistinguishable, non-interacting particles.

## 8.1 Equilibrium steady state after Bose-Einstein condensation

The equilibrium steady state of the Bose gas in Eq. (6.2) is entirely defined by the steady state number distribution $p_N(N_0, T) \equiv p_N(N_0, t \to \infty)$ of Eq. (6.3). We introduce the abbreviation $\lambda^{\pm}_{\leftrightarrows}(N - N_0, T) \to \lambda^{\pm}_{\leftrightarrows}(N_0, T)$ in the following. Due to Eq. (6.5), the net flux of particles between condensate and non-condensate is zero[1], $\partial \langle N_0 \rangle / \partial t = 0$, if the condensate number distribution obeys the recursion relation:

$$p_N(N_0 + 1, T) = p_N(N_0, T) \frac{\lambda^+_{\leftrightarrows}(N_0, T)}{\lambda^-_{\leftrightarrows}(N_0 + 1, T)} , \qquad (8.1)$$

---
[1]This is the case, if the non-condensate particles in the gas reach the critical density, i.e. $\varrho_\perp \lambda^3 = \zeta(3/2)$ for a gas of $N$ particles in a box.

# Chapter 8. EQUILIBRIUM PROPERTIES OF A DILUTE BOSE-EINSTEIN CONDENSATE

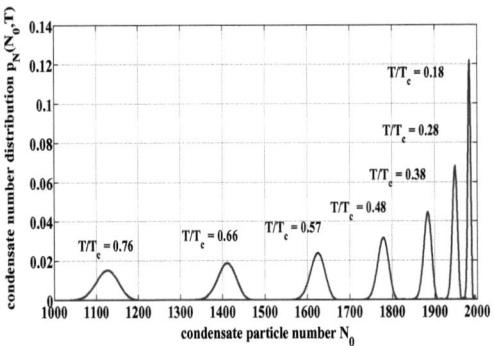

Figure 8.1: *Steady state condensate particle number distribution $p_N(N_0, T)$ in Eq. (8.2) for $N = 2000$ $^{87}$Rb atoms in a three-dimensional harmonic trap with frequencies $\omega_x = \omega_y = 2\pi \times 42.0$ Hz, $\omega_z = 2\pi \times 120.0$ Hz, for different temperatures $T/T_c = 0.76, 0.66, 0.57, 0.48, 0.38, 0.28, 0.17$, given an ideal gas critical temperature $T_c = 33.86$ nK. The distribution $p_N(N_0, T)$ unambiguously describes the non-condensate number distribution by replacing $N_0 \to N - N_0$ in the above figure.*

which leads to the steady state distribution

$$p_N(N_0, T) = \mathcal{N} \prod_{z=1}^{N_0-1} \frac{\lambda^+_{\leadsto}(z-1, T)}{\lambda^-_{\leadsto}(z, T)}, \tag{8.2}$$

with normalization $\mathcal{N} = \sum_{N_0=0}^{N} \prod_{z=1}^{N_0} \lambda^+_{\leadsto}(z-1, T)/\lambda^-_{\leadsto}(z, T)$. Since Eq. (6.3) cannot adopt multiple steady states, a proof of the uniqueness of the steady state in Eq. (8.2) can be found in Appendix A.4. Note that the steady state distribution in Eq. (8.2) implies $\partial_t \langle N_0 \rangle = 0$ and therefore energetic equilibrium and detailed balanced particle flow (on average), i.e. $\lambda^+_{\leadsto}(\langle N_0 \rangle, T) = \lambda^-_{\leadsto}(\langle N_0 \rangle, T)$ and $\mu_0 = \mu_\perp(N - \langle N_0 \rangle, T)$ following from Eq. (6.6) – in agreement with thermodynamics. Typical condensate particle number distributions are shown in Fig. 8.1 for $N = 2000$ atoms in a three-dimensional harmonic trap with frequencies $\omega_x = \omega_y = 2\pi \times 42.0$ Hz, $\omega_z = 2\pi \times 120.0$ Hz, for different temperatures.

## 8.2 On the quantum ergodicity conjecture

Ergodicity is a statistical assumption made originally by Ludwig Boltzmann in 1872 for a classical gas of non-interacting particles, and means that a gas is supposed to reach a unique equilibrium state (a thermal state) after long times, where each available state of the same energy is sampled equally (in classical phase space) over time. The ergodicity conjecture is often used to simplify theoretical treatments of interacting many particle quantum systems, using a thermal state ansatz for the Bose gas below its critical temperature, and neglecting atomic interactions in sufficiently weakly interacting, dilute atomic gases.

In order to prove this ergodicity conjecture for a weakly interacting Bose-Einstein condensate (undergoing Markovian dynamics), we are left to show that the unique and stable steady state of the Bose gas defined by Eqs. (6.2, 8.2) is a Boltzmann state. Hence, our steady state is compared in the perturbative limit of weak interactions to a thermal state of $N$ non-interacting particles at temperature $T$,

$$\hat{\sigma}_{N,\text{th}} = \hat{\mathcal{Q}}_N \frac{e^{-\beta \hat{\mathcal{H}}}}{\mathcal{Z}(N,T)} \hat{\mathcal{Q}}_N , \qquad (8.3)$$

with the partition function of $N$ indistinguishable particles represented by $\mathcal{Z}(N,T) = \text{Tr}\{\hat{\sigma}_{N,\text{th}}\}$. Again, $\hat{\mathcal{Q}}_N$ is the projector onto the Fock space of $N$ particles. In the absence of interactions, $\hat{\mathcal{H}} = \sum_{\vec{k}} \eta_{\vec{k}} \hat{a}^{\dagger}_{\vec{k}} \hat{a}_{\vec{k}}$ in Eq. (8.3) denotes the Hamiltonian of an ideal gas, thus refering to Eq. (3.4) with $g \equiv 0$. To prove the equality of the state $\hat{\sigma}_{N,\text{th}}$ and the steady state of the Bose gas in Eqs. (6.2, 8.2) arising from the master equation, it is to be shown that the following exact recursion relation (which was derived from Eq. (8.3) in Section 1.5.2) for the condensate particle number distribution,

$$\frac{p_{N,\text{th}}(N_0, T)}{p_{N,\text{th}}(N_0 + 1, T)} = e^{\beta \eta_0} \frac{\mathcal{Z}_{\perp}(N - N_0, T)}{\mathcal{Z}_{\perp}(N - N_0 - 1, T)} , \qquad (8.4)$$

# Chapter 8. EQUILIBRIUM PROPERTIES OF A DILUTE BOSE-EINSTEIN CONDENSATE

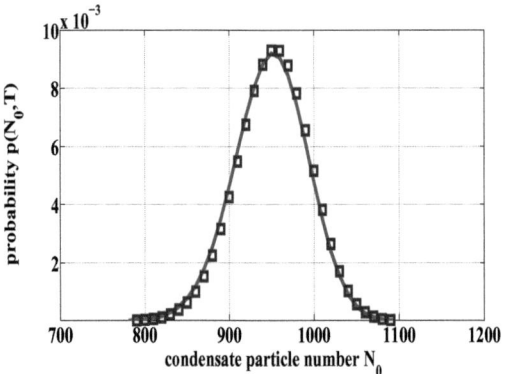

Figure 8.2: Comparison of exact condensate particle number distribution of the master equation (red solid line) vs. the condensate statistics arising from Boltzmann thermal state for an ideal quantum gas in Eq. (8.3) (blue squares), for $N = 2000$ atoms in a three-dimensional harmonic trap with frequencies $\omega_x = \omega_y = 2\pi \times 42.0$ Hz and $\omega_z = 2\pi \times 120.0$ Hz. The gas temperature is $T = 25.0$ nK. Similar agreement is observed for different temperatures.

applies for the steady state of the master equation in Eq. (8.2) in the formal limiting case of weak interactions, $a\varrho^{1/3} \to 0^+$ with $a \neq 0$. In Eq. (8.4), $\mathscr{Z}_\perp(N - N_0)$ denotes the partition function of $(N - N_0)$ non-condensate particles, see Eq. (5.13), and $\eta_0$ is the single particle ground state energy of a non-interacting gas. The analytical proof is figured out straight forwardly, because only basic elements of the master equation formalism developed in Chapters 3-8 are to be employed. Approximating $\lambda^-_{\leadsto}(N_0, T) \approx \lambda^-_{\leadsto}(N_0 - 1, T)$ and therewith neglecting terms of the order of $N^{-1}$, the steady state solution of the condensate particle number distribution of the master equation, Eq. (8.2), leads to:

$$\frac{p_N(N_0, T)}{p_N(N_0+1, T)} \simeq \frac{\lambda^-_{\leadsto}(N_0, T)}{\lambda^+_{\leadsto}(N_0, T)} = e^{\beta(\eta_0 - \mu_\perp(N-N_0, T))} , \qquad (8.5)$$

where we used (i) the balance condition between condensate feeding and losses rates in Eq. (7.20) which applies for two body interactions exhibiting finite spatial

## 8.2. ON THE QUANTUM ERGODICITY CONJECTURE

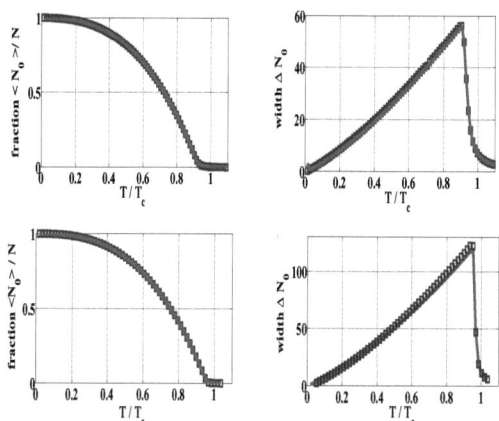

Figure 8.3: (color online) Average condensate fraction $\langle N_0 \rangle(t)/N$ and standard deviation $\Delta N_0$ of the condensate particle number distribution $p_N(N_0, T)$, obtained from the steady state distribution (red solid line) of the master equation vs. the canonical ensemble prediction (blue squares) of Eq. (8.3), for trap frequencies of $\omega_x = \omega_y = 42.0$ Hz and $\omega_z = 120.0$ Hz, and atom numbers of $N = 2000$ and $N = 10000$ atoms. The gases have corresponding critical temperatures of $T_c = 33.86$ nK (upper figures), and $T_c = 57.90$ nK (lower figures).

phase coherence time, $\beta \hbar \Gamma \ll 1$, and (ii) that $\eta_0 = \mu_0$ in the formal limit of small interactions, $a\varrho^{1/3} \to 0^+$ (see Section 7.5). The non-condensate chemical potential, as defined by the normalization condition in Eq. (7.8) is related [10] to the non-condensate partition function $\mathscr{Z}_\perp(N - N_0, T)$ in Eq. (5.13), due to $\mu_\perp(N - N_0, T) = -\beta^{-1}[\ln \mathscr{Z}_\perp(N - N_0, T) - \ln \mathscr{Z}_\perp(N - N_0 - 1, T)]$, see Eq. (7.19) in Section 7.2. Therewith, we arrive at the recurrence relation in Eq. (8.4).

Hence, the steady state of the entire Bose gas in Eq. (6.2) is given by the thermal state of an ideal gas projected onto the subspace of $N$ particles in Eq. (8.3), in the formal limit of weak interactions, $a\varrho^{1/3} \to 0^+$ (with $a \neq 0$). Remembering and comparing again the two figures 3.2 and 1.4, this is by no means trivial, since the steady state distribution in Eq. (8.2) a priori depends on the (specific nonlinearity of the) two body interaction term $\hat{V}_{0\perp}$ according to the feeding and loss rates $\lambda^\pm_\leftrightarrows(N_0, T)$, see Eqs. (7.3). Nevertheless, the balances of particle flow between the different single particle eigenmodes of the gas at equilibrium generate the same statistics as a thermal state of an ideal gas (with $a \equiv 0$) projected onto the subspace of $N$ particles in Eq. (8.3). In both cases, the condensate number statistics thus obeys the distribution

$$p_N(N_0, T) = e^{-\beta \eta_0 N_0} \frac{\mathscr{Z}_\perp(N - N_0, T)}{\mathscr{Z}(N, T)}, \tag{8.6}$$

where $\mathscr{Z}_\perp(N - N_0, T)$ is the non-condensate partition function in Eq. (5.13), and $\mathscr{Z}(N, T)$ the partition function of the canonical ensemble in Eq. (1.26). Obviously, the condensate number distribution is modified with respect to the steady state solution of the master equation for a harmonic oscillator coupled to a heat bath by the term $\mathscr{Z}_\perp(N - N_0, T)/\mathscr{Z}(N, T)$ below $T_c$.

The quantitative support of the $(1/N)$ approximation required for the analytical proof is to compare the exact numerical calculations of the steady state condensate particle number distribution to the prediction of the Boltzmann ansatz in Eqs. (8.3). Figure 8.2 verifies that for $N = 2000$ atoms in a three-dimensional harmonic trap with frequencies $\omega_x = \omega_y = 2\pi \times 42.0$ Hz and $\omega_z = 2\pi \times 120.0$ Hz, and a gas temperature $T = 20.0$ nK, the exact numerical calculations (red solid line) agree almost perfectly with the Boltzmann thermal state (blue squares).

In Fig. 8.3, the comparison of the average condensate occupations $\langle N_0 \rangle$ and the width $\Delta N_0$ of the condensate particle number distributions is displayed as a function of the entire range of relative temperatures $T/T_c$ for $N = 2000$ and $N = 10000$ atoms, given the same trap parameters as in Fig. 8.2. The corresponding critical temperatures are $T_c = 33.86$ nK and $T_c = 57.90$ nK. Agreement is observed between the steady state of the master equation and the Boltzmann ansatz: The shift of the critical temperature is about 10% with respect to the ideal gas critical temperature $T_c$ (in the semiclassical limit) in both cases, and maxima of the standard deviations are $\Delta N_0 = 50.4$ at $T = 0.90 T_c$ for $N = 2000$ atoms, and $\Delta N_0 = 124.71$ at $T = 0.96 T_c$ for

$N = 10000$. The thermal state ansatz is hence supported by numerical calculations, which have been reproduced and checked for different available parameter ranges of state-of-the-art experiments.

## 8.3 Exact condensate statistics versus semiclassical limit

The condensate statistics obtained in the semiclassical limit is contrasted to the predictions for gases with discrete spectra. For this purpose, the equilibrium steady state in Eq. (8.2) is appropriate to compare condensate number fluctuations and average occupations in the Bose gas in a three-dimensional harmonic trap.

### 8.3.1 Condensate particle number distribution

The steady state solution of Eq. (6.3) for an interacting Bose gas in a 3-dimensional harmonic trap with trapping frequencies $\omega_x, \omega_y$ and $\omega_z$ is formally given by

$$p_N(N_0, T) = \mathcal{N} \prod_{z=0}^{N_0} \exp[\beta \Delta \mu (N - z)] \,, \qquad (8.7)$$

with normalization $\mathcal{N} = \sum_{N_0=0}^{N} \prod_{z=0}^{N_0} \exp[\beta \Delta \mu (N - z)]$.

Remember that Eq. (8.7) defines the equilibrium steady state of the full $N$-body state $\hat{\sigma}^{(N)}(\infty)$, which – for a three-dimensional harmonic trap – has the explicit form

$$\hat{\sigma}^{(N)}(\infty) = p_N(N_0, T) |N_0\rangle\langle N_0| \otimes \sum_{\{N_{\vec{k}}\}}^{(N-N_0)} p_N(\{N_{\vec{k}}\}, T) |\{N_{\vec{k}}\}\rangle\langle\{N_{\vec{k}}\}| \,, \qquad (8.8)$$

where the distribution of non-condensate occupations is given by

$$p_N(\{N_{\vec{k}}\}, T) = \mathscr{L}_\perp^{-1}(N - N_0, T) \prod_{\vec{k}} e^{-\beta N_{\vec{k}} \epsilon_{\vec{k}}} \,. \qquad (8.9)$$

180    Chapter 8. EQUILIBRIUM PROPERTIES OF A DILUTE BOSE-EINSTEIN CONDENSATE

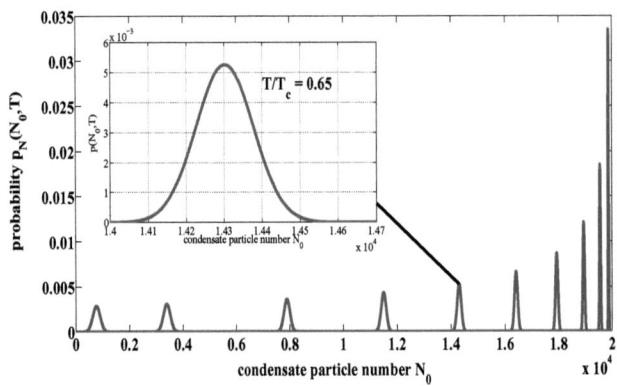

Figure 8.4: *Condensate particle number distribution $p_N(N_0, T)$ for $N = 20000$ atoms in a three-dimensional harmonic trap with frequencies $\omega_x = \omega_y = 42.0$ Hz, and $\omega_z = 120.0$ Hz, for different temperatures ranging from $T/T_c = 0.99, 0.94, 0.85, 0.75, 0.65, 0.55, 0.47, 0.38, 0.28, 0.20$ nK (from left to right). The ideal gas critical temperature is $T_c = 72.94$ nK. Inset displays the magnification of $p_N(N_0, T = 0.64 T_c)$.*

In Eq. (8.9), $\mathscr{Z}_\perp(N-N_0, T)$ constitutes the partition function of $(N-N_0)$ non-condensate particles in Eq. (5.13), and $\epsilon_{\vec{k}}$ denotes the single particle energies of non-condensate particles in Eq. (4.29).

Numerically studying the perturbative limit of weakly interacting gases, $a\varrho^{1/3} \to 0^+$, single particle energies turn into the ones of a non-interacting Bose gas, $\epsilon_{\vec{k}} \to \eta_{\vec{k}}$, while $\Delta\mu(N-z, T) \to \mu_\perp(N-z, T) - \eta_{\vec{0}}$. The non-condensate chemical potential is thus specified by the following normalization condition for the non-condensate particle number,

$$\sum_{\vec{k} \neq \vec{0}} \frac{1}{e^{\beta(\eta_{\vec{k}} - \mu_\perp(N-z,T))} - 1} = (N - z) , \qquad (8.10)$$

which determines $\mu_\perp(N-z, T)$ in a numerically exact way, for any $z$. For a three-dimensional harmonic trap, the invariance of the statistics under the shift of the single particle ground state energy can be employed, using $\eta_{\vec{k}} = (k_x \hbar \omega_x + k_y \hbar \omega_y +$

## 8.3. EXACT CONDENSATE STATISTICS VERSUS SEMICLASSICAL LIMIT

$k_z\hbar\omega_z$) as the single particle energies and setting $\eta_0 = 0$. Expanding the exponential function in Eq. (8.10) and using the harmonic series [40] leads to

$$(N-z) = \sum_{j=1}^{\infty} \exp[j\beta\mu_\perp(N-z,T)]\left[\prod_{\xi=x,y,z}\frac{1}{1-\exp[-j\beta\hbar\omega_\xi]} - 1\right]. \quad (8.11)$$

Equations (8.7) and (8.11) present an accessible numerical tool in order to exactly calculate the condensate particle number distribution: For each $z = 0\ldots N$, the implicit Eq. (8.11) for $\mu_\perp(N-z,T)$ is determined numerically, which directly yields the condensate particle number distribution $p_N(N_0,T)$ in Eq. (8.7). The distribution is shown in Fig. 8.4, for a Bose-Einstein condensate of $N = 20000$ atoms in a three-dimensional harmonic trap with frequencies $\omega_x = \omega_y = 2\pi \times 42.0$ Hz, $\omega_z = 2\pi \times 120.0$ Hz, for different temperatures $T/T_c$, and an ideal gas critical temperature $T_c = 72.94$ nK.

Now, we pay attention to the semiclassical approximation. The semiclassical limit is useful to analytically deduce the scaling behavior for the moments of the distribution $p_N(N_0,T)$ (see Section 8.4). Within this limit, the approximate steady state distribution is obtained [15] by replacing the summation in Eq. (8.10) by an integration over the density of states, $g(\eta) = \eta^2(2\hbar\omega_x\omega_y\omega_z)^{-1}$. In that case, the non-condensate chemical potential, and therewith the condensate particle number distribution in Eq. (8.7) is defined by the implicit equation

$$m_3[e^{\beta\mu_\perp(N-z,T)}] = (N-z)\frac{\hbar^3\omega_x\omega_y\omega_z}{(k_BT)^3}, \quad (8.12)$$

with $m_3[z] = \sum_{k=1}^{\infty} z^k/k^3$, the Bose function for a three-dimensional harmonic trap [15].

Although the replacement of the summation by an integration is often employed in order to derive analytical predictions, it should be emphasized that it is not exact (because it miscounts the number of single particle states), which leads to a shift of the critical temperature. Hence this shift is not induced by the neglect

# Chapter 8. EQUILIBRIUM PROPERTIES OF A DILUTE BOSE-EINSTEIN CONDENSATE

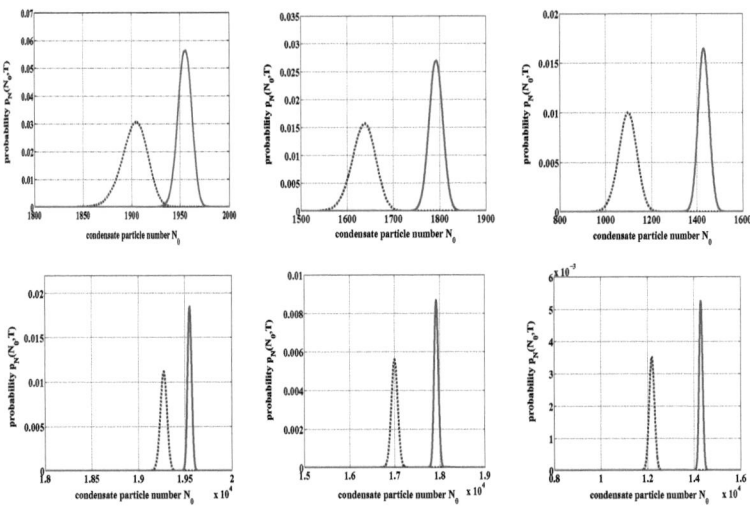

Figure 8.5: *Comparison of condensate particle number distributions $p_N(N_0, T)$ obtained from the exact quantum calculation in Eq. (8.7, 8.11) (dashed blue line) vs. the semiclassical limit in Eqs. (8.7, 8.12) (solid red line), for the same trap parameters as in Fig. 8.4, and $N = 2000$ and $N = 20000$ particles in the trap. Relative temperatures are $T/T_c = 0.3, 0.5$ and $0.7$ (from left to right), with corresponding critical temperatures of $T_c = 33.86$ nK for $N = 2000$, and $T_c = 72.94$ for $N = 20000$ atoms.*

of the zero-point motion [15].[2]. To explicitly illustrate these deviations originating from the semiclassical approximation, the exact condensate particle number distribution in Eqs. (8.7, 8.11) (blue dotted curves) is compared to the distribution in the semiclassical limit in Eqs. (8.7, 8.12) (red solid lines) in Fig. 8.5. Calculations are performed for a harmonic trap with frequencies $\omega_x = \omega_y = 2\pi \times 42.0$ Hz, $\omega_z = 2\pi \times 120.0$ Hz, once for $N = 2000$ (top), and once for $N = 20000$ atoms (bottom) in the trap, with corresponding ideal gas critical temperatures $T_c = 33.86$ nK and $T_c = 72.94$ nK. Three different temperatures, $T/T_c = 0.3, 0.5$ and $0.7$, are displayed from left to right.

For $N = 2000$ particles, the maxima of the condensate particle number distribution

---

[2]The shift of the critical temperature follows the analytical law given in Eq. (8.13), see Section 8.3.3, which is derived under the inclusion of the exact single particle spectrum.

## 8.3. EXACT CONDENSATE STATISTICS VERSUS SEMICLASSICAL LIMIT

$p_N(N_0, T)$ occur at $N_0 = 1956$, $N_0 = 1793$ and $N_0 = 1430$, using the semiclassical limit of Eqs. (8.7, 8.12), whereas they are located at $N_0 = 1905$, $N_0 = 1638$ and $N_0 = 1097$, within the exact quantum calculation with Eq. (8.7, 8.11). For $N = 20000$ particles, the maxima are found at $N_0 = 19550$, $N_0 = 17920$ and $N_0 = 14280$ with the distribution in the semiclassical limit in Eqs. (8.7, 8.12), whereas the maxima are located at $N_0 = 19270$, $N_0 = 17000$ and $N_0 = 12180$ within the exact quantum distribution governed by Eqs. (8.7, 8.11). Hence, even though deviations have a vanishing trend as compared to the total number of particles $N$ in the trap, the distributions (maxima and widths) significantly differ relatively up to 2 – 30%, even for large total particle numbers of $N = 20000$ atoms in the trap.

### 8.3.2 Average condensate occupation and number variance

The resulting shift of the critical temperature is best deduced from the analysis of the condensate number expectation value, and its standard deviation. In Fig. 8.6, the average condensate occupation $\langle N_0 \rangle$, and corresponding standard deviations $\Delta N_0$ of the particle number distribution $p_N(N_0, T)$ in the semiclassical limit (blue diamonds) represented by Eqs. (8.7, 8.12) are compared to the exact quantum calculation via Eqs. (8.7, 8.11) (red squares). Herefore, we use a continuous range of relative temperatures $T/T_c > \hbar\omega_i/k_B T_c$, and two different total number of atoms, $N = 2000$ (left panels) and $N = 10000$ (right panels), in a three-dimensional harmonic trap with trapping frequencies $\omega_x = \omega_y = 42.0$ Hz, $\omega_z = 120.0$ Hz. The corresponding ideal gas critical temperatures in the semiclassical limit are $T_c = 33.86$ nK and $T_c = 57.90$ nK.

For a total number of $N = 2000$ atoms (left panels), the quantum calculation leads to a shift of the critical temperature of about 10% with respect to the ideal gas critical temperature $T_c$ in the semiclassical limit in Eq. (1.21): The maximum width depicts the critical point of the phase transition. It is $\Delta N_0 = 50.42$ occuring at $T = 0.90 T_c$ for the quantum calculation, whereas we observe $\Delta N_0 = 38.96$ at $T = T_c$ in the semiclassical limit. For larger total atom numbers, here $N = 10000$ (right

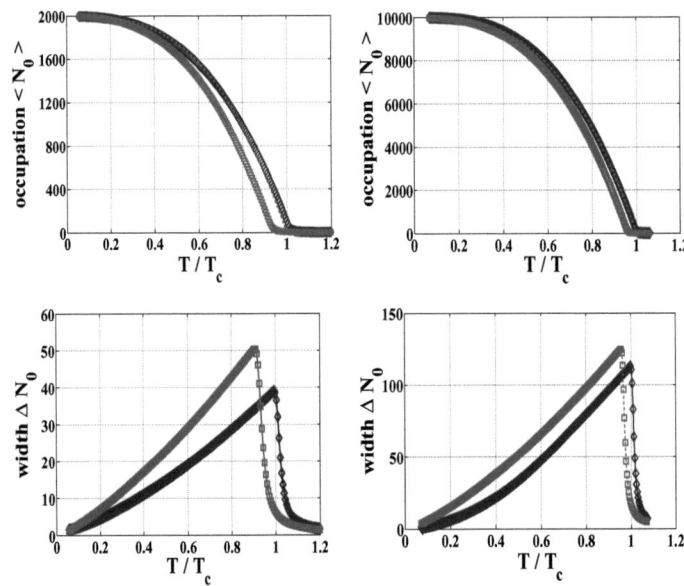

Figure 8.6: *Average condensate occupations and standard deviation of the steady state distribution $p_N(N_0, T)$ in the semiclassical limit of Eqs. (8.7, 8.12) (blue diamonds), $N \to \infty$ and $k_B T \gg \hbar\omega_i$, vs. exact calculation, using Eqs. (8.7, 8.11) (red squares). Trap parameters are $\omega_x = \omega_y = 42.0$ Hz, $\omega_z = 120.0$ Hz, with atom numbers $N = 2000$ and $N = 10000$, and corresponding critical temperatures $T_c = 33.86$ nK, and $T_c = 57.90$ nK.*

figures), the critical temperature of the exact calculation is $0.96 T_c$ still 4-5% less than $T_c$ – even for the case of relatively large particle numbers. Maximum widths are $\Delta N_0 = 124.71$ at $T = 0.96 T_c$ for the quantum calculation, whereas $\Delta N_0 = 112.81$ at $T = T_c$ in the semiclassical limit.

### 8.3.3 Shift of the critical temperature

Even though both calculations (exact vs. semiclassical approximation) follow the same qualitative trend, i.e., condensate particle number fluctuations getting maximal at the critical point and condensate occupations following the typical $N(1-$

$T^3/T_c^3$) scaling behavior (see Chapter 1), the critical temperature of the exact calculation is shifted significantly with respect to the semiclassical ansatz for mesoscopic Bose gases. These deviations have their origin in the neglect of the degeneracy of the discrete single particle spectrum [80]. Getting pronounced ($\sim 10-30\%$) for low particle numbers ($N \sim 10^3$), as likely to be used in recent experiments [12], deviations still occur for relatively large atomic samples ($N \sim 10^4$), leading to corrections of $4-5\%$ to the ideal gas critical temperature $T_c$ in the semiclassical limit.

An analytical estimate (including the exact single particle spectrum) for the shift of the critical temperature [80] agreeing with our numerical results is given by

$$\frac{T_c^{\text{exact}}}{T_c} = 1 - \frac{\zeta(2)\sum_{\eta=x,y,z}\omega_\eta}{6\zeta(3)^{2/3}N^{1/3}\bar{\omega}}, \qquad (8.13)$$

where $T_c$ is the critical temperature of an ideal gas in the thermodynamic limit, and $\bar{\omega} = (\omega_x\omega_y\omega_z)^{1/3}$ the averaged trap frequency. The shift of the exact critical temperature $T_c^{\text{exact}}$ to the ideal gas value $T$ is typically of the order of $2-30\%$ for state-of-the-art experimental parameters, and can only be neglected if $N \to \infty$, where the large number of atoms leading to large single particle occupation turns to be equivalent to the approximation of the single particle spectrum to be a continuous one.

## 8.4 Analytical scaling behaviors in the semiclassical limit

Finally, the moments of the condensate particle number distribution $p_N(N_0, T)$ are analytically studied in the semiclassical limit, becoming quantitatively accurate in the limit of large particle numbers, $N \to \infty$.

### 8.4.1 Condensate and non-condensate particle number distribution

In the semiclassical limit, $\exp[\beta\Delta\mu_\perp(z, N-z, T)]$ can be assumed to approach unity in Eqs. (8.7, 8.12). Thus, using the approximate relation $m_3[\exp[\beta\Delta\mu_\perp(z, N-z, T)]] \simeq$

$\zeta(3)\exp[\beta\Delta\mu_\perp(z, N-z, T)]$ in Eq. (8.12) and employing that $T_c^3 = \hbar^3\omega_x\omega_y\omega_z/(k_B T)^3 N$ in the semiclassical limit, turns Eq. (8.7) to the normalized condensate and non-condensate number distribution

$$p_N(N_0, T) = p_N(N - N_0, T) \approx e^{-\lambda} \frac{\lambda^{(N-N_0)}}{(N-N_0)!} \frac{N!}{\Gamma(N+1, \lambda)}, \quad (8.14)$$

valid for $T < T_c$ in the semiclassical limit. In Eq. (8.14), $\lambda = NT^3/T_c^3$ is the mean number of non-condensate particles, and

$$\Gamma(N+1, \lambda) = \int_\lambda^\infty dt \, t^N e^{-t} = N! \, e^\lambda \sum_{k=0}^N \frac{\lambda^k}{k!} \quad (8.15)$$

is an incomplete Gamma function [40] related to the finite particle number in the gas. As the particle number is considered to be large, $N \to \infty$, the incomplete Gamma function $\Gamma(N+1, \lambda) \to 1$, and thus approaches unity.

Equation (8.14) discloses that the non-condensate particle number distribution $p_N(N - N_0, T)$ is Poissonian in the number of non-condensate particles, and is thus distributed around the average non-condensate occupation $\lambda$. Its width $\Delta N_\perp$ scales as the root of the average non-condensate particle number $\Delta N_\perp = \sqrt{\lambda}$. The Poisson distribution of the non-condensate particle number highlights the statistical independence of the non-condensate particles at final thermal equilibrium.

In contrast, the condensate particle distribution, $p_N(N_0, T)$, is clearly not Poissonian, meaning that the particle number is distributed around the average condensate occupation $\langle N_0 \rangle = N - \lambda$, and the width of the condensate particle number distribution is given by $\Delta N_0 = \sqrt{\lambda}$. This again reflects that the condensate statistics is entirely defined by the non-condensate part of the gas, the width $\Delta N_0$ being proportional to the number of thermal non-condensate atoms at equilibrium. Thus, the smaller the temperature, the higher the number of coherent atoms, and the smaller the variance in the condensate particle number.

## 8.4. ANALYTICAL SCALING BEHAVIORS IN THE SEMICLASSICAL LIMIT

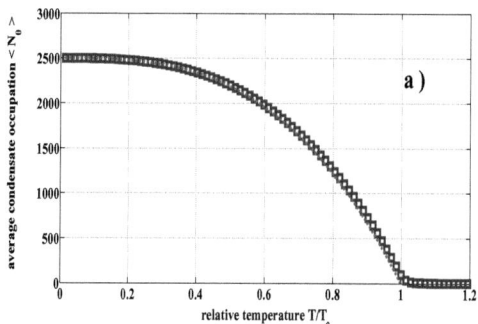

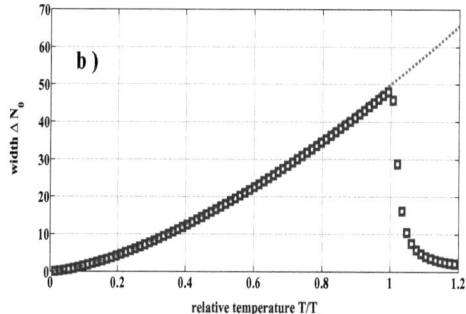

Figure 8.7: *Scaling behavior in the semiclassical limit. Panel a) compares average condensate occupation $\langle N_0 \rangle$ as a function of relative temperature $T/T_c$ of the numerically obtained condensate particle number distribution $p_N(N_0, T)$ in Eqs. (8.7, 8.12) (blue squares) to the analytical result $\langle N_0 \rangle = N(1 - T^3/T_c^3)$ in Eq. (8.14) (dahes red line), for a gas of $N = 2500$ atoms with identical trapping parameters as in Fig. 8.4. The ideal gas critical temperature is $T_c = 36.47$ nK. Figure b) compares the standard deviation $\Delta N_0$ obtained numerically from Eqs. (8.7, 8.12) (blue squares) to the analytical prediction $\Delta N_0 = NT^{3/2}/T_c^{3/2}$ of Eq. (8.14) (dashed red line).*

## 8.4.2 Average condensate occupation and number variance

The average condensate ground state occupation $\langle N_0 \rangle$ is studied in the semiclassical limit as a function of the relative temperature $T/T_c$. The ground state occupation is shown in Fig. 8.7a for a gas of $N = 2500$ atoms in a three-dimensional harmonic trap with trapping frequencies $\omega_x = \omega_y = 2\pi \times 42.0$ Hz, $\omega_z = 2\pi \times 120.0$ Hz, following the analytical prediction arising from Eq. (8.14):

$$\langle N_0 \rangle = N \left( 1 - \left( \frac{T}{T_c} \right)^3 \right) . \tag{8.16}$$

The scaling behavior of $\langle N_0 \rangle$ in Fig. (8.7) is universal for different trapping parameters and particle numbers. Due to particle number conservation, it follows that the average occupation number of non-condensate particles is given by

$$\langle (N - N_0) \rangle = N T^3 / T_c^3 . \tag{8.17}$$

The standard deviation $\Delta N_0(T)$ of the condensate particle number distribution as a function of relative temperature $T/T_c$ predicted by the (semiclassical limit) steady state $p_N(N_0, T)$ of the master equation in Eqs. (8.7, 8.12) is displayed in Fig. 8.7b (blue squares) for the same parameters as in Fig. 8.7a: Condensate particle number fluctuations get maximal exactly at the critical point of an ideal gas in the semiclassical limit, $T = T_c$, a universal behavior which doesn't qualitatively vary for different trap and gas parameters. The scaling behavior of the condensate and non-condensate particle number, $\Delta N_0(T) = \Delta N_\perp(T)$, again follows from Eq. (8.14):

$$\Delta N_\perp = \Delta N_0 = \sqrt{N \left( \frac{T}{T_c} \right)^3} . \tag{8.18}$$

The analytical result of Eq. (8.18) is shown in Fig. 8.7b (red dashed line), resembling the numerical result obtained from the semiclassical distribution $p_N(N_0, T)$ in

## 8.4. ANALYTICAL SCALING BEHAVIORS IN THE SEMICLASSICAL LIMIT

Eqs. (8.7, 8.12) (blue squares). Note again that the scaling behavior in the semiclassical limit in Eqs. (8.16, 8.17, 8.18) is only valid in the limit of continuous single particle spectra, or in the limit of large particle numbers $N \to \infty$ (see Section 8.3). Indeed, the distribution in Eq. (8.14) applies only for $T < T_c$.

### 8.4.3 Higher order moments of the steady state distribution

The approximately given condensate particle number distribution $p_N(N_0, T)$ in Eq. (8.14) furthermore specifies all central moments of the condensate and the non-condensate particle number distribution $p_N(N - N_0, T) = p_N(N_0, T)$ analytically. Calculations of these moments are typically defined by technically involved Bell or Touchard polynomials [27]. Here, they are obtained by approximating the Poissonian non-condensate particle number distribution in Eq. (8.14) by a Gaussian distribution, which is reasonable for sufficiently large particle numbers. Thereby, we get the Gaussian non-condensate particle number distribution $p_N(N - N_0, T)$:

$$p_N(N_0, T) = p_N(N - N_0, T) \approx \frac{1}{\sqrt{2\pi\lambda}} e^{\frac{-(N-N_0-\lambda)^2}{2\lambda}}, \qquad (8.19)$$

with a mean value of $N - N_0$, and a variance of the non-condensate particle number $\langle (N - N_0)^2 \rangle$, equal to the average non-condensate particle number $\lambda = NT^3/T_c^3$. The Gaussian ansatz in Eq. (8.19) yields in particular a Gaussian approximation for the condensate particle number distribution $p_N(N_0, T)$ by replacing $(N - N_0) \to N_0$ in Eq. (8.19), as $p_N(N_0, T) = p_N(N - N_0, T)$.

Now, the $n^{th} = (2k)^{th}$ central moment of the non-condensate particle number distribution is analytically given [40, 27] by

$$\langle (N - N_0 - \lambda)^{2k} \rangle = \frac{(2k)!}{2^k k!} \lambda^k, \qquad (8.20)$$

for even $n$, and is zero otherwise. Indeed, due to particle number conservation, all central moments $\langle (N_0 - \langle N_0 \rangle)^n \rangle$ of the condensate particle number distribution

# Chapter 8. EQUILIBRIUM PROPERTIES OF A DILUTE BOSE-EINSTEIN CONDENSATE

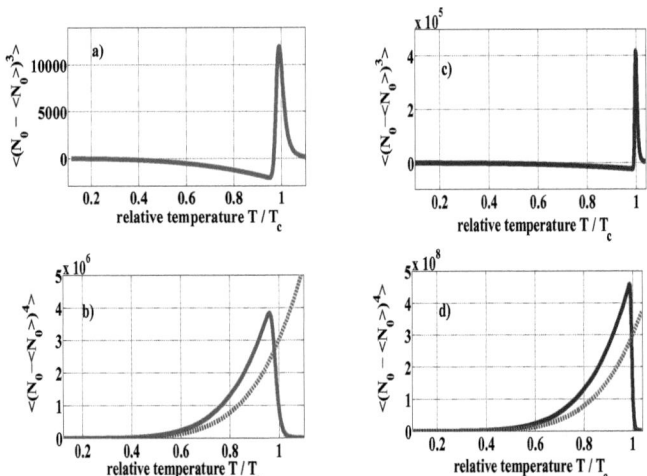

Figure 8.8: *Third- and fourth-order moments of the condensate particle number distribution $p_N(N_0, T)$ in Eqs. (8.7, 8.11), for a gas of $N = 1000$ (red solid lines, figures a) and b)), and $N = 10000$ (blue solid lines, figures c) and d)) atoms in a three-dimensional trap with identical trap parameters as in Fig. 8.4, and hence critical temperatures of $T_c = 26.87$ nK and $T_c = 57.90$ nK. Dashed lines in figures b) and d) correspond to the analytical prediction $\langle (N_0 - \langle N_0 \rangle)^4 \rangle = 3N^2 T^6/T_c^6$ of the Gaussian ansatz in Eq. (8.19), whereas the third central moment of the Gaussian distribution is zero everywhere.*

$p_N(N_0, T)$ are also fully specified [27] by Eq. (8.19), because

$$\langle (N_0 - \langle N_0 \rangle)^{2k} \rangle = \langle (N - N_0 - \lambda)^{2k} \rangle \equiv (2k)!(2^k k!)^{-1} \lambda^k \ . \tag{8.21}$$

Comparing the first four central moments according to Eq. (8.19) to numerical calculations via Eqs. (8.7, 8.12) in Figs. 8.7 demonstrates that the mean value and the variance of the Gaussian ansatz exactly equals $\lambda$, following the analytical prediction in Eq. (8.14). The numerically obtained third moment in Figs. 8.8a and Figs. 8.8c, however, exhibits a non-trivial behavior close $T = T_c$, where the distribution gets increasingly asymmetric in the vicinity of the phase transition. In contrast, the Gaussian third-order moment vanishes for all temperatures. We con-

## 8.4. ANALYTICAL SCALING BEHAVIORS IN THE SEMICLASSICAL LIMIT

clude that the Gaussian prediction of $\langle (N_0 - \langle N_0 \rangle)^3 \rangle = 0$ matches the numerically obtained third order moment for a gas of $N = 1000$ atoms as shown in Fig. 8.8a for temperatures $T_c = 0\ldots0.5$, whereas for larger particle numbers, $N = 10000$ in Fig. 8.8c, an agreement is observed for $T/T_c = 0\ldots0.7$.

The fourth order moments in Figs. 8.8b and 8.8d scale comparably to the analytical prediction of the Gauss approximation $\langle (N_0 - \langle N_0 \rangle)^4 \rangle = N^2 T^6 / T_c^6$ up to $T/T_c = 0.6$ for both values of the total particle number, i.e. for $N = 1000$ and for $N = 10000$ atoms.

The first four moments predicted by the Gaussian ansatz are hence either valid for almost all temperatures below the critical point, if the particle number is sufficiently large, e.g. up to $T/T_c = 0.7$ for $N \sim 10000$, or for sufficiently small temperatures, e.g. up to $T/T_c = 0.4$ for rather small particle numbers $N \sim 1000$. We conjecture that the Gaussian ansatz becomes exact in the limit $N \to \infty$.

Chapter 9

# Final conclusions

## 9.1 Master equation of Bose-Einstein condensation

**Condensate formation**
Our central conceptual result is the Markov master equation of Bose-Einstein condensation in Eq. (6.3), which explains the time evolution of the full $N$-body state $\hat{\sigma}^{(N)}(t)$ of the gas undergoing the Bose-Einstein phase transition. Our equation describes Bose-Einstein condensation in terms of two body collisions, takes into account the depletion of the non-condensate, avoids a state factorization into a condensate and a non-condensate density matrix and models the non-equilibrium number statistics during condensate formation. Identifying the small parameter $a\varrho^{1/3}$ for dilute atomic gases in the condensate number transition rates, we could numerically monitor the entire condensate number distribution during Bose-Einstein condensation for the first time. It requires the calculation of $2(N+1)$ single particle feeding and loss rates, and a numerical solution procedure to solve for the $(N+1)$ coupled differential equations, all together executable on standard serial computers.

**Equilibrium steady state of a Bose-Einstein condensate**
We have derived a unique steady state for the $N$-body state of the Bose gas. Using

the limit of dilute, weakly interacting atomic gases and employing the Markovian dynamics assumption, we could show that the steady state of a Bose-Einstein condensate is unique and stable. In the limit of weak interactions, this steady state is in particular given by a Boltzmann thermal state of an ideal gas, projected onto the subspace of exactly $N$ indistinguishable particles.

**Time scales for condensate formation**
The condensate formation times predicted by our theory match the correct order of magnitude of experimentally observed time scales for condensate formation. We were able to estimate the order of magnitude for intrinsic energy shifts in a Bose-Einstein condensate, confirming the conjecture [17, 18] that they are small for dilute and mesoscopic atomic gases.

## 9.2 What is Bose-Einstein condensation?

Thermodynamically, we see Bose-Einstein condensation as a relaxation process of the gas below $T_c$ until the Boltzmann entropy is maximized and the free energy is minimized, according to the (generalized) Einstein de Broglie condition.

Quantum mechanically, the Bose-Einstein phase transition is marked as such by *strong number fluctuations* which set in when the coherence length $\lambda(T)$ of the non-condensate particles exceeds their average distance. The time evolution of the condensate number distribution during condensate formation highlights the full $N$-body quantum dynamics of the particles during condensate formation, while the wave picture implies that all particles overlap below $T_c$.

Quantum statistically, condensate (and non-condensate) number distributions in the final *equilibrium steady state* of a dilute, weakly interacting Bose-Einstein condensate (undergoing Markovian dynamics) are uniquely captured by a *Boltzmann thermal state of an ideal gas*, subject to the statistics of $N$ indistinguishable particles.

## 9.3 Outlook

Our master equation theory describes in particular the many particle dynamics after sudden deformations of the trap geometry, which remains to be analyzed in detail. In addition, an extension to study the dynamics induced by an additional external electromagnetic field is possible. We have reformulated the theory for two-component spinor Bose-Einstein condensates, which is of our future interest [97, 98].

Possible improvements of the presented master equation consist in taking into account the finite thermalization time of the non-condensate during Bose-Einstein condensation, in particular using a microscopic first principles derivation of the decay rate $\Gamma$ between condensate and non-condensate correlations. The inclusion of pair processes for studying condensate formation and for deriving the steady state of the gas may modify quantitative predictions, if the single particle ground state energy exceeds the energies of non-condensate single particle states (e.g. for quasi one-dimensional Bose gases [99, 100, 101, 102], where in particular the geometry dependence of the scattering amplitude may lead to physically different scenarios [102]). A detailed numerical comparison of unperturbed to perturbed transition rates including the time dependence of the condensate mode for single particle, pair and scattering processes is also planned to be carried out in future works.

# APPENDIX

# Appendix A

# Important proofs and calculations

## A.1 Correlation functions of the non-condensate field

The normally and anti-normally ordered two point correlation functions of the non-condensate field are decomposed for single particle ($\rightsquigarrow$), pair ($\leftrightsquigarrow$) and scattering ($\circlearrowright$) processes, using Wick's theorem [94]. We begin with the normally ordered two point correlation function for single particle processes, $\mathcal{G}^{(+)}_{\rightsquigarrow}(\vec{r},\vec{r}',N-N_0,T,\tau)$ in Eq. (5.32):

$$\mathcal{G}^{(+)}_{\rightsquigarrow}(\vec{r},\vec{r}',N-N_0,T,\tau) = \left\langle \hat{\Psi}^{\dagger}_{\perp}(\vec{r},\tau)\hat{\Psi}^{\dagger}_{\perp}(\vec{r},\tau)\hat{\Psi}_{\perp}(\vec{r},\tau)\hat{\Psi}^{\dagger}_{\perp}(\vec{r}',0)\hat{\Psi}^{\dagger}_{\perp}(\vec{r}',0)\hat{\Psi}_{\perp}(\vec{r}',0) \right\rangle^{(N-N_0)}_{\mathscr{F}_{\perp}} = \\ 2\left\langle \hat{\Psi}^{\dagger}_{\perp}(\vec{r},\tau)\hat{\Psi}_{\perp}(\vec{r}',0) \right\rangle^{(N-N_0)}_{\mathscr{F}_{\perp}} \left\langle \hat{\Psi}^{\dagger}_{\perp}(\vec{r},\tau)\hat{\Psi}_{\perp}(\vec{r}',0) \right\rangle^{(N-N_0)}_{\mathscr{F}_{\perp}} \left\langle \hat{\Psi}_{\perp}(\vec{r},\tau)\hat{\Psi}^{\dagger}_{\perp}(\vec{r}',0) \right\rangle^{(N-N_0)}_{\mathscr{F}_{\perp}}. \quad \text{(A.1)}$$

The anti-normally ordered correlation function for single particle processes, $\mathcal{G}^{(-)}_{\rightsquigarrow}(\vec{r},\vec{r}',N-N_0,T,\tau)$ in Eq. (5.33), can be decomposed similarly:

$$\mathcal{G}^{(-)}_{\rightsquigarrow}(\vec{r},\vec{r}',N-N_0,T,\tau) = \left\langle \hat{\Psi}^{\dagger}_{\perp}(\vec{r},\tau)\hat{\Psi}_{\perp}(\vec{r},\tau)\hat{\Psi}_{\perp}(\vec{r},\tau)\hat{\Psi}^{\dagger}_{\perp}(\vec{r}',0)\hat{\Psi}^{\dagger}_{\perp}(\vec{r}',0)\hat{\Psi}_{\perp}(\vec{r}',0) \right\rangle^{(N-N_0)}_{\mathscr{F}_{\perp}} = \\ 2\left\langle \hat{\Psi}^{\dagger}_{\perp}(\vec{r},\tau)\hat{\Psi}_{\perp}(\vec{r}',0) \right\rangle^{(N-N_0)}_{\mathscr{F}_{\perp}} \left\langle \hat{\Psi}_{\perp}(\vec{r},\tau)\hat{\Psi}^{\dagger}_{\perp}(\vec{r}',0) \right\rangle^{(N-N_0)}_{\mathscr{F}_{\perp}} \left\langle \hat{\Psi}_{\perp}(\vec{r},\tau)\hat{\Psi}^{\dagger}_{\perp}(\vec{r}',0) \right\rangle^{(N-N_0)}_{\mathscr{F}_{\perp}}. \quad \text{(A.2)}$$

The non-condensate field $\hat{\Psi}_\perp(\vec{r}, \tau)$ in the interaction picture with respect to $\hat{\mathcal{H}}_\perp$ in Eq. (4.26), written in the single particle basis set $\{|\Psi_k\rangle, k \in \mathbb{N}\}$ of the non-condensate, turns into

$$\hat{\Psi}_\perp(\vec{r}, \tau) = \hat{\mathcal{U}}_\perp(\tau)\hat{\Psi}_\perp(\vec{r})\hat{\mathcal{U}}_\perp^\dagger(\tau) = \sum_{k \neq 0} \Psi_k(\vec{r})\hat{a}_k \exp\left[-\frac{i\epsilon_k \tau}{\hbar}\right] . \qquad (A.3)$$

Any two point correation function of products of two non-condensate fields in Eqs. (A.1, A.2) can hence be written as a function of the average occupation of different non-condensate single particle states $|\Psi_k\rangle \in \mathscr{F}_\perp$:

$$\left\langle \hat{\Psi}_\perp^\dagger(\vec{r}, \tau)\hat{\Psi}_\perp(\vec{r}', 0) \right\rangle_{\mathscr{F}_\perp}^{(N-N_0)} = \sum_{k \neq 0} \Psi_k^*(\vec{r})\Psi_k(\vec{r}')f_k(N-N_0, T) \exp\left[-\frac{i\epsilon_k \tau}{\hbar}\right] , \qquad (A.4)$$

where

$$\left\langle \hat{a}_k^\dagger \hat{a}_l \right\rangle_{\mathscr{F}_\perp}^{(N-N_0)} = \left\langle \hat{a}_k^\dagger \hat{a}_k \right\rangle_{\mathscr{F}_\perp}^{(N-N_0)} \delta_{kl} \equiv f_k(N-N_0, T)\delta_{kl} . \qquad (A.5)$$

The function

$$f_k(N-N_0, T) = \left\{ \hat{a}_k^\dagger \hat{a}_k \hat{\mathscr{D}}_{N-N_0} \frac{e^{-\beta \hat{\mathcal{H}}_\perp}}{\mathscr{Z}_\perp(N-N_0)} \hat{\mathscr{D}}_{N-N_0} \right\} \qquad (A.6)$$

describes the average many particle occupation of a non-condensate single particle state $|\Psi_k\rangle$, given that $(N - N_0)$ particles are in the non-condensate. An explicit analytical derivation for the expressions of the occupation numbers $f_k(N-N_0, T)$ is given in Appendix A.3.

Anti-normal products of two point correlation functions of two non-condensate fields in the interaction picture arising in Eqs. (A.1, A.2) can be obtained correspondingly, i.e., they turn into

## A.1. CORRELATION FUNCTIONS OF THE NON-CONDENSATE FIELD

$$\left\langle \hat{\Psi}_\perp(\vec{r},\tau)\hat{\Psi}_\perp^\dagger(\vec{r}',0) \right\rangle_{\mathscr{F}_\perp}^{(N-N_0)} = \sum_{k\neq 0} \Psi_k(\vec{r})\Psi_k^*(\vec{r}')[f_k(N-N_0,T)+1]\,\exp\left[\frac{i\epsilon_k\tau}{\hbar}\right], \quad (A.7)$$

given that

$$\left\langle \hat{a}_k \hat{a}_l^\dagger \right\rangle_{\mathscr{F}_\perp}^{(N-N_0)} = [f_k(N-N_0,T)+1]\delta_{kl}. \quad (A.8)$$

With respect to single particle processes, we hence find for normally and anti-normally ordered two point correlation functions:

$$\mathscr{G}^{(+)}(\vec{r},\vec{r}',N-N_0,T,\tau) = 2\sum_{k,l,m\neq 0} \Psi_k^*(\vec{r})\Psi_k(\vec{r}')\Psi_l^*(\vec{r}')\Psi_l(\vec{r}')\Psi_m^*(\vec{r})\Psi_m(\vec{r}')[f_k(N-N_0,T)+1] \times$$
$$\times f_l(N-N_0,T)f_m(N-N_0,T)\,\exp\left[\frac{i(\epsilon_k-\epsilon_l-\epsilon_m)\tau}{\hbar}\right], \quad (A.9)$$

$$\mathscr{G}^{(-)}(\vec{r},\vec{r}',N-N_0,T,\tau) = 2\sum_{k,l,m\neq 0} \Psi_k^*(\vec{r})\Psi_k(\vec{r}')\Psi_l^*(\vec{r})\Psi_l(\vec{r}')\Psi_m^*(\vec{r})\Psi_m(\vec{r}')f_k(N-N_0,T) \times$$
$$\times [f_l(N-N_0,T)+1][f_m(N-N_0,T)+1]\,\exp\left[\frac{-i(\epsilon_k-\epsilon_l-\epsilon_m)\tau}{\hbar}\right]. \quad (A.10)$$

Integration of $\mathscr{G}^{(\pm)}(\vec{r},\vec{r}',N-N_0,T,\tau)\exp[-\tau^2/\tau_{\text{col}}^2]$ over the time interval $\tau$, multiplied by $\Psi_0(\vec{r})\Psi_0^*(\vec{r}')\exp[\pm i\omega_0\tau]$, which arises from the backswitch of the condensate fields from the interaction picture, $\hat{\Psi}_0(\vec{r},\tau) \simeq \hat{\Psi}_0(\vec{r})\exp[\pm i\omega_0\tau]$ using the Gross-Pitaevskii equation (4.4), leads to the single particle loss and feeding rates in Eq. (7.3).

Through the decomposition of a product of four non-condensate fields into a product of two time-ordered two point correlation functions of two non-condensate fields in the two point correlation functions for pair events, $\mathscr{G}^{(\pm)}(\vec{r},\vec{r}',N-N_0,T,\tau)$, and applying Eq. (A.7, A.8), the normally ordered correlation function for pair events turns into:

$$\begin{aligned}\mathscr{G}^{(+)}_{\sim\sim}(\vec{r},\vec{r}',N-N_0,T,\tau) &= \left\langle \hat{\Psi}^\dagger_\perp(\vec{r},\tau)\hat{\Psi}^\dagger_\perp(\vec{r}',\tau)\hat{\Psi}_\perp(\vec{r},0)\hat{\Psi}_\perp(\vec{r}',0)\right\rangle^{(N-N_0)}_{\mathscr{F}_\perp} = \\ &\quad 2\left\langle \hat{\Psi}^\dagger_\perp(\vec{r},\tau)\hat{\Psi}_\perp(\vec{r}',0)\right\rangle^{(N-N_0)}_{\mathscr{F}_\perp}\left\langle \hat{\Psi}^\dagger_\perp(\vec{r}',\tau)\hat{\Psi}_\perp(\vec{r}',0)\right\rangle^{(N-N_0)}_{\mathscr{F}_\perp} = \\ &\quad 2\sum_{k,l\neq 0}\Psi^*_k(\vec{r})\Psi_k(\vec{r}')\Psi^*_l(\vec{r})\Psi_l(\vec{r}')f_k(N-N_0,T)f_l(N-N_0,T)\exp\left[\frac{-i(\epsilon_k+\epsilon_l)\tau}{\hbar}\right].\end{aligned} \quad (A.11)$$

The anti-normally ordered pair correlation function $\mathscr{G}^{(-)}_{\sim\sim}(\vec{r},\vec{r}',N-N_0,T,\tau)$ can be decomposed similarly:

$$\begin{aligned}\mathscr{G}^{(-)}_{\sim\sim}(\vec{r},\vec{r}',N-N_0,T,\tau) &= \left\langle \hat{\Psi}_\perp(\vec{r},\tau)\hat{\Psi}_\perp(\vec{r}',\tau)\hat{\Psi}^\dagger_\perp(\vec{r},0)\hat{\Psi}^\dagger_\perp(\vec{r}',0)\right\rangle^{(N-N_0)}_{\mathscr{F}_\perp} = \\ &\quad 2\left\langle \hat{\Psi}_\perp(\vec{r},\tau)\hat{\Psi}^\dagger_\perp(\vec{r}',0)\right\rangle^{(N-N_0)}_{\mathscr{F}_\perp}\left\langle \hat{\Psi}_\perp(\vec{r}',\tau)\hat{\Psi}^\dagger_\perp(\vec{r}',0)\right\rangle^{(N-N_0)}_{\mathscr{F}_\perp} = \\ &\quad 2\sum_{k,l\neq 0}\Psi^*_k(\vec{r})\Psi_k(\vec{r}')\Psi^*_l(\vec{r})\Psi_l(\vec{r}')[f_k(N-N_0,T)+1][f_l(N-N_0,T)+1]\exp\left[\frac{-i(\epsilon_k+\epsilon_l)\tau}{\hbar}\right],\end{aligned} \quad (A.12)$$

which after multiplication with $\Psi^2_0(\vec{r})\left(\Psi^*_0(\vec{r}')\right)^2\exp[\pm 2i\omega_0\tau]$ and integration over $\tau$ turns into the pair feeding and loss rates in Eq. (7.12).

Finally, the scattering two point correlation function $\mathscr{G}_\circlearrowleft(\vec{r},\vec{r}',N-N_0,T,\tau)$ is decomposed from a product of four non-condensate fields into a product of two time ordered two point correlation functions of two non-condensate fields:

$$\begin{aligned}\mathscr{G}_\circlearrowleft(\vec{r},\vec{r}',N-N_0,T,\tau) &= \left\langle \hat{\Psi}^\dagger_\perp(\vec{r},\tau)\hat{\Psi}_\perp(\vec{r},\tau)\hat{\Psi}^\dagger_\perp(\vec{r}',0)\hat{\Psi}_\perp(\vec{r}',0)\right\rangle^{(N-N_0)}_{\mathscr{F}_\perp} = \\ &\quad \left\langle \hat{\Psi}^\dagger_\perp(\vec{r},\tau)\hat{\Psi}_\perp(\vec{r}',0)\right\rangle^{(N-N_0)}_{\mathscr{F}_\perp}\left\langle \hat{\Psi}_\perp(\vec{r},\tau)\hat{\Psi}^\dagger_\perp(\vec{r}',0)\right\rangle^{(N-N_0)}_{\mathscr{F}_\perp} = \\ &\quad 2\sum_{k,l\neq 0}\Psi^*_k(\vec{r})\Psi_k(\vec{r}')\Psi^*_l(\vec{r})\Psi_l(\vec{r}')f_k(N-N_0,T)[f_l(N-N_0,T)+1]\exp\left[\frac{i(\epsilon_k-\epsilon_l)\tau}{\hbar}\right],\end{aligned} \quad (A.13)$$

which, after multiplication with $|\Psi_0(\vec{r})|^2|\Psi_0(\vec{r}')|^2$ and integration over $\tau$ turns into the

## A.2 Detailed balance conditions

scattering rate in Eq. (7.16).

## A.2 Detailed balance conditions

We proof the balance condition

$$\lambda_+^\sim(N-N_0,T) = \exp[\beta\Delta\mu(N-N_0,T)]\lambda_-^\sim(N-N_0,T) \tag{A.14}$$

between single particle feedings and losses, where $\Delta\mu(N-N_0,T) = \mu_\perp(N-N_0,T) - \mu_0$ is the difference between the eigenvalue of the Gross-Pitaevskii equation $\mu_0$ in Eq. (4.4), and $\mu_\perp(N-N_0,T)$ represents the non-condensate chemical potential in Eq. (A.28). Since the single particle body feeding and loss rates are given by

$$\lambda_\pm^\sim(N-N_0,T) = \frac{8\pi^3\hbar^2 a^2}{m^2} \sum_{k,l,m \neq 0} \mathscr{W}_\pm^\sim(k,l,m,N-N_0,T)\delta^{(\Gamma)}(\omega_k+\omega_l-\omega_m-\omega_0), \tag{A.15}$$

where the $\delta$-distribution is given by Eq. (7.4), it is sufficient to show that

$$\mathscr{W}_+^\sim(k,l,m,N-N_0,T) = \exp[\beta\Delta\mu(N-N_0,T)]\mathscr{W}_-^\sim(k,l,m,N-N_0,T) \tag{A.16}$$

under the constraint that $\omega_k + \omega_l - \omega_m + \omega_0 \leq \Delta$. Since, as a matter of fact, the probability amplitudes satisfy $(\zeta_{kl}^{m0})^* = \zeta_{m0}^{kl}$, the exact relation $[f_k(N-N_0,T)+1] = f_k(N-N_0,T)\exp[\beta(\epsilon_k - \mu_\perp(N-N_0,T))]$ for the occupation numbers $f_k(N-N_0,T)$ of non-condensate single particle states in Eq. (A.24) further enables us to show that

$$\begin{aligned}\mathscr{W}_+^\sim(k,l,m,N-N_0,T) &= f_k(N-N_0,T)f_l(N-N_0,T)[f_m(N-N_0,T)+1]|\zeta_{kl}^{m0}|^2 \\ &= f_k(N-N_0,T)f_l(N-N_0,T)f_m(N-N_0,T)\exp[\beta(\epsilon_m - \mu_\perp(N-N_0,T))]|\zeta_{kl}^{m0}|^2 \\ &= f_k(N-N_0,T)f_l(N-N_0,T)f_m(N-N_0,T)\exp[\beta(\epsilon_k+\epsilon_l - \mu_0 - \mu_\perp(N-N_0,T) - \hbar\Delta)]|\zeta_{m0}^{kl}|^2 \\ &= [f_k(N-N_0,T)+1][f_l(N-N_0,T)+1]f_m(N-N_0,T)\exp[\beta(\mu_\perp(N-N_0,T)-\mu_0)]|\zeta_{m0}^{kl}|^2 \\ &= \exp[\beta\Delta\mu(N-N_0,T)]\mathscr{W}_-^\sim(k,l,m,N-N_0,T), \end{aligned} \tag{A.17}$$

with $\Delta\mu = \mu_\perp(N - N_0, T) - \mu_0$. Because of the finite width $\sim \Gamma$ of the $\delta$-function in Eq. (A.15), the balance condition in Eq. (A.14) is valid for $\beta\hbar\Gamma \ll 1$. This is the case in the parameter regime of dilute gases, where we checked that $\equiv \exp[-\beta\hbar\Gamma] \simeq 1$ for $\beta\hbar\Gamma = \beta\hbar\sqrt{2}/\tau_{col} \sim 10^{-3} \ll 1$.

## A.3  Occupation numbers of the non-condensate

The state of the non-condensate in Eq. (5.14) allows to determine the average number of particles, $f_k = f_k(N - N_0, T)$ in Eq. (A.6), in each particular non-condensate single particle mode $|\Psi_k\rangle$, given that $N_0$ particles populate the condensate mode and $(N - N_0)$ particles the non-condensate single particle modes. According to Eq. (A.6), the expectation value of the number operator $\hat{N}_k$ in a non-condensate state of $(N - N_0)$ particles is

$$f_k(N - N_0, T) = \mathscr{Z}_\perp^{-1}(N - N_0) \sum_{\{N_k\}}^{(N-N_0)} N_k \exp\left[-\beta \sum_{k \neq 0} \epsilon N_k\right], \quad (A.18)$$

where $\mathscr{Z}_\perp(N - N_0)$ is the partition function of $(N - N_0)$ indistinguishable particles in the non-condenste in Eq. (5.15). In terms of the partial partition sum, $\mathscr{Z}_\perp^{(k)}(N - N_0)$ [10], which excludes the sum over the particular mode $|\Psi_k\rangle$, Eq. (A.24) can be rewritten as

$$f_k(N - N_0, T) = \mathscr{Z}_\perp^{-1}(N - N_0) \sum_{N_k=0}^{(N-N_0)} N_k \exp[-\beta \epsilon_k N_k] \mathscr{Z}_\perp^{(k)}(N - N_0 - N_k) \ . \quad (A.19)$$

For small enough $N_k$ (it suffices to start at $N_k = 1$ and to determine $\mathscr{Z}_\perp^{(k)}(N - N_0 - N_k)$ iteratively), it is possible to expand

$$\log\left[\mathscr{Z}_\perp^{(k)}(N - N_0 - 1)\right] \simeq \log\left[\mathscr{Z}_\perp^{(k)}(N - N_0)\right] - \alpha_\perp^{(k)}(N - N_0, T) \ , \quad (A.20)$$

## A.3. OCCUPATION NUMBERS OF THE NON-CONDENSATE

with the parameter

$$\alpha_\perp^{(k)}(N-N_0,T) = \frac{\partial \log\left[\mathscr{L}_\perp^{(k)}(N-N_0)\right]}{\partial(N-N_0)}. \tag{A.21}$$

From Eq. (A.20), we find the recursion relation

$$\frac{\mathscr{L}_\perp^{(k)}(N-N_0-1)}{\mathscr{L}_\perp^{(k)}(N-N_0)} = \exp\left[-\alpha_\perp^{(k)}(N-N_0,T)\right] \tag{A.22}$$

between the partial partition sums $\mathscr{L}_\perp^{(k)}(N-N_0)$ of $N-N_0$, and $\mathscr{L}_\perp^{(k)}(N-N_0-1)$ of $N-N_0-1$ non-condensate particles. Multiple iteration of Eq. (A.22) leads to

$$\frac{\mathscr{L}_\perp^{(k)}(N-N_0-N_k)}{\mathscr{L}_\perp^{(k)}(N-N_0)} = \exp\left[-N_k \alpha_\perp^{(k)}(N-N_0,T)\right], \tag{A.23}$$

and Eq. (A.19) turns into

$$f_k(N-N_0,T) = \frac{\mathscr{L}_\perp^{(k)}(N-N_0)}{\mathscr{L}_\perp(N-N_0)} \sum_{N_k=0}^{(N-N_0)} N_k \exp\left[-\left(\beta\epsilon_k + \alpha_\perp^{(k)}(N-N_0,T)\right)N_k\right]. \tag{A.24}$$

It remains to apply the same procedure to the partition function $\mathscr{L}_\perp(N-N_0)$. Using the decomposition in Eq. (A.19), and applying Eq. (A.23), one finds that

$$\mathscr{L}_\perp(N-N_0) = \mathscr{L}_\perp^{(k)}(N-N_0) \sum_{N_k=0}^{(N-N_0)} \exp\left[-\left(\beta\epsilon_k + \alpha_\perp^{(k)}(N-N_0,T)\right)N_k\right]. \tag{A.25}$$

Setting Eq. (A.25) into Eq. (A.24), the expectation value of particle number occupations of a particular non-condensate single particle state $|\Psi_k\rangle$, given that $(N-N_0)$

particles populate the non-condensate modes, turns into

$$f_k(N-N_0,T) = \frac{1}{\exp\left[\beta\epsilon_k + \alpha_\perp^{(k)}(N-N_0,T)\right]-1} . \quad (A.26)$$

Hence, the parameter $\alpha^{(k)}$ can be interpreted as to describe the change in temperature in the non-condensate part of the gas during condensate formation. The parameter $\alpha_\perp^{(k)}(N-N_0,T)$ is approximately independent of the state $k$ [10], i.e. the change in temperature during condensation is described by one single parameter, $\alpha^{(k)} \simeq \alpha_\perp(N-N_0,T)$. The latter is determined by the constraint of particle number conservation, as spelled out by the implicit equation

$$\sum_{k\neq 0} f_k(N-N_0,T) = \sum_{k\neq 0} \frac{1}{\exp[\beta\epsilon_k + \alpha_\perp(N-N_0,T)]-1} = (N-N_0) . \quad (A.27)$$

As evident from Eq. (A.21), and from the fact that each subspace of $(N-N_0)$ particles is in a thermal state (of microscopic occupation number corresponding to different temperatures), the parameter $\alpha_\perp(N-N_0,T)$ can be interpreted as the ratio of the non-condensate chemical potential for a state of $(N-N_0)$ atoms to the thermal energy $\beta^{-1}$ [10]. Hence, from the definition in Eq. (A.21), we see that $\alpha_\perp(N-N_0,T)$ is, upon a constant, nothing more than the derivative of the Helmholtz free energy $\mathcal{F}(N-N_0) = -\beta^{-1}\log\mathcal{Z}_\perp(N-N_0)$ of the $(N-N_0)$ particles in the non-condensate, thus given [10] by

$$\alpha_\perp(N-N_0,T) = -\beta\frac{\partial \mathcal{F}}{\partial(N-N_0)} = -\beta\mu_\perp(N-N_0,T) , \quad (A.28)$$

introducing the non-condensate chemical potential $\mu_\perp(N-N_0,T)$.

## A.4 Proof of uniqueness of the Bose gas' steady state

We proof the uniqueness of the equilibrium steady state of the Bose gas, defined by Eqs. (6.2) and (8.2). To this end, it is to show that

$$p_N(N_0, T) = \mathcal{N} \prod_{z=1}^{N_0-1} \frac{\lambda^{(+)}_{\leadsto}(z-1, T)}{\lambda^{(-)}_{\leadsto}(z, T)} \quad \Leftrightarrow \quad \frac{\partial p_N(N_0, t)}{\partial t} = 0 , \quad (A.29)$$

"$\Rightarrow$": First, let's verify that $p_N(N_0, T)$ in Eq. (8.2) is a stationary solution of Eq. (6.3). Assume $p_N(N_0, T)$ to be given by Eq. (8.2). Equation (6.3) then implies that

$$p_N(N_0 + 1) = p_N(N_0) \frac{\lambda^{+}_{\leadsto}(N_0)}{\lambda^{-}_{\leadsto}(N_0 + 1, T)} , \quad (A.30)$$

which leads to a vanishing of the terms in Eq. (6.3) proportional to $N_0$:

$$N_0 p_N(N_0 - 1, T)\lambda^{(+)}_{\leadsto}(N_0 + 1, T) - N_0 p_N(N_0, T)\lambda^{(-)}_{\leadsto}(N_0, T) = 0 . \quad (A.31)$$

Moreover, Eq. (A.30) shows that the same applies to the terms in Eq. (6.3) proportional to $N_0 + 1$:

$$(N_0 + 1)p_N(N_0 + 1, T)\lambda^{(-)}_{\leadsto}(N_0 + 1, T) - (N_0 + 1)p_N(N_0, T)\lambda^{(+)}_{\leadsto}(N_0, T) = 0 . \quad (A.32)$$

Therefore, the distribution $p_N(N_0, T)$ given by Eq. (8.2) is a stationary solution of the evolution Eq. (6.3):

$$p_N(N_0, T) = \mathcal{N} \prod_{z=1}^{N_0-1} \frac{\lambda^{(+)}_{\leadsto}(z-1, T)}{\lambda^{(-)}_{\leadsto}(z, T)} \quad \Rightarrow \quad \frac{\partial p_N(N_0, T)}{\partial t} = 0 . \quad (A.33)$$

"$\Leftarrow$": Let's now prove that $p_N(N_0, T)$ in Eq. (A.29) is the *unique* solution of Eq. (6.3).

Suppose that $\partial_t p_N(N_0, T) = 0$. By induction, it can be proven that the recurrence relation for the steady state distribution $p_N(N_0, T)$ arising from Eq. (6.3) equals Eq. (A.30) for all $N_0$. From Eq. (6.3), it follows that

$$p_N(N_0+1, T) = \frac{\left(N_0 \lambda_\leadsto^{(-)}(N_0, T) + (N_0+1)\lambda_\leadsto^{(+)}(N_0, T)\right) p_N(N_0, T) - N_0 p_N(N_0-1, T)\lambda_\leadsto^{(+)}(N_0+1, T)}{(N_0+1)\lambda_\leadsto^{(-)}(N_0+1, T)} .$$
(A.34)

Choosing the starting point of induction at $N_0 = 1$ ($\mathcal{N}$ may be chosen arbitrarily, hence it was set to one), it follows from Eq. (A.34) that

$$p_N(1, T) = \lambda_\leadsto^+(0, T)/\lambda_\leadsto^-(1, T) ,$$
(A.35)

which conincides with Eq. (A.30), for $N_0 = 1$.

The induction step is $N_0 \to N_0 + 1$: Suppose Eqs. (A.34) and (A.30) equal for $N_0$. Let's write Eq. (A.34) for arbitrary $N_0 + 1$:

$$p_N(N_0+2, T) = \frac{\left[(N_0+1)\lambda_\leadsto^{(-)}(N_0+1, T) + (N_0+2)\lambda_\leadsto^{(+)}(N_0+1, T)\right] p_N(N_0+1, T)}{(N_0+2)\lambda_\leadsto^{(-)}(N_0+2, T)}$$

$$- \frac{(N_0+1)\lambda_\leadsto^{(+)}(N_0, T) p_N(N_0, T)}{(N_0+2)\lambda_\leadsto^{(-)}(N_0+2, T)}$$
(A.36)

which turns into

$$p(N_0+2, T) = \frac{\lambda_\leadsto^{(+)}(N_0+1, T)}{\lambda_\leadsto^{(-)}(N_0+2, T)} p_N(N_0+1, T) ,$$
(A.37)

under the use of the induction assumption in Eq. (A.30) for $N_0$. Equation A.37 equals Eq. (A.30) for $N_0+1$ condensate particles, hence proofing our statement that $p_N(N_0, T)$ in Eq. (8.2) is the unique equilibrium steady distribution of Eq. (6.3). The

## A.5. NON-CONDENSATE THERMALIZATION

unique $N$-body equilibrium steady state is therefore given by

$$\hat{\sigma}^{(N)}(t \to \infty) = \sum_{N_0=0}^{N} p_N(N_0, T)|N_0\rangle\langle N_0| \otimes \hat{\rho}_\perp(N - N_0, T) \tag{A.38}$$

after long times with

$$p_N(N_0, T) = \mathcal{N} \prod_{z=1}^{N_0-1} \frac{\lambda_{\leadsto}^{(+)}(z-1, T)}{\lambda_{\leadsto}^{(-)}(z, T)} . \tag{A.39}$$

As shown in Chapter 6, this steady state turns into a Gibbs-Boltzmann thermal state of an ideal gas for sufficiently small interactions captured by the formal limiting case $a\varrho^{1/3} \to 0^+$.

## A.5 Non-condensate thermalization

We treat the backgroud gas as a thermalized, depleted thermal gas, which is equivalent to assume the non-condensate thermalization to be ideally infinitely fast as compared to the condensate formation time. Despite the fact that this has been explicitly demonstrated in the experiment [64], and previous theoretical approaches [60, 64, 76, 51], we shall also convince ourselves numerically that this assumption is well satisfied for harmonic traps. An estimate for the non-condensate thermalization time in the presence of a condensate is given [82, 47] by:

$$\frac{1}{\tau_\perp} = 6.47 * \left(\frac{a}{L}\right)^{7/5} N^{17/30} \left(\frac{T}{T_c}\right)^{1/2} \left(\frac{N_0}{N}\right)^{2/5} \bar{\omega}, \tag{A.40}$$

where $L = (\hbar/m\bar{\omega})^{1/2}$ is the extension of the harmonic oscillator ground state, $\bar{\omega} = (\omega_x \omega_y \omega_z)^{1/3}$ is the averaged frequency of an anisotropic trap, and $T_c = (\hbar\bar{\omega}/k_B)[N/\zeta(3)]^{1/3}$ the critical temperature of an ideal gas. In Fig. A.1, we show the comparison of the time scales for non-condensate relaxation $\tau_\perp \sim \tau_{col}$ to typical condensate

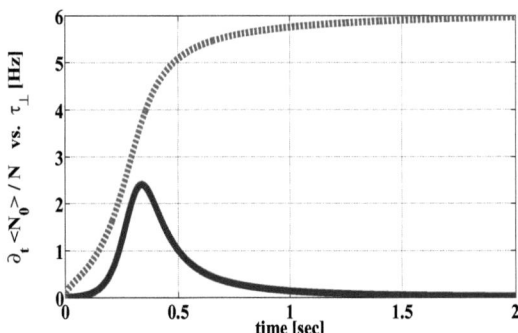

Figure A.1: *Probing the rapid non-condensate thermalization in a shock cooling process. For sufficiently smooth condensate time evolution, the assumption of rapid non-condensate equilibration is well satisfied, since the thermalization in the non-condensate is faster than the change of the condensate fraction: Figure shows the comparison of the thermal redistribution rate $\tau_\perp$ [82, 47] (red dashed line) to condensate flux $\partial_t \langle N_0 \rangle / N$ (blue solid line) as a function of time for $N = 2000$ $^{87}$Rb atoms in the gas with temperature $T = 30.0$ nK. The critical temperature is $T_c = 53.07$ nK, for an isotropic trap with frequencies $2\pi v_x = 2\pi \cdot v_y = 2\pi \cdot v_z = 600$ Hz.*

formation times $\tau_0$ obtained from Eq. (6.6): The assumption of rapid non-condensate thermalization is satisfied initially, still holds approximately at the exponential stage of condensate growth (and is well satisfied, if the condensate growth is smooth), and is very well satisfied again at the final stage of condensate formation.

# Bibliography

[1] R. Onofrio, C. Raman, J. M. Vogels, J. R. Abo-Shaeer, A. P. Chikkatur, and W. Ketterle. Observation of superfluid flow in a Bose-Einstein condensed gas. *Phys. Rev. Lett.*, 85:2228, (2000).

[2] A. J. Leggett. Bose-Einstein condensation in the alkali gases: Some fundamental concepts. *Rev. Mod. Phys.*, 73:307, (2001).

[3] A. Fetter. Vortices in an imperfect Bose gas I: The condensate. *Phys. Rev.*, 138:A429–A437, (1965).

[4] F. Chevy, K. W. Madison, and J. Dalibard. Measurement of the angular momentum of a rotating Bose-Einstein condensate. *Phys. Rev. Lett.*, 85:2223, (2000).

[5] P. Engels, I. Coddington, P. C. Haljan, and E. A. Cornell. Non-equilibrium effects of anisotropic compression applied to vortex lattices in Bose-Einstein condensates. *Phys. Rev. Lett.*, 89:100403, (2002).

[6] P. W. Anderson. Absence of diffusion in certain random lattices. *Phys. Rev.*, 109:1492, (1958).

[7] S. E. Skipetrov, A. Minguzzi, B. A. van Tiggelen, and B. Shapiro. Anderson localization of a Bose-Einstein condensate in a 3D random potential. *Phys. Rev. Lett.*, 100:165301, (2008).

[8] S. Schelle, D. Delande, and A. Buchleitner. Mircowave driven atoms: From

Anderson localization to Einstein's photo effect. *Phys. Rev. Lett.*, 102:183001, (2009).

[9] M. Albiez, R. Gati, J. Fölling, B. Hemmerling, M. Cristiani, and M. Oberthaler. Direct observation of tunneling and nonlinear self-trapping in a single bosonic Josephson junction. *Phys. Rev. Lett.*, 95:010402, (2005).

[10] F. Reif. *Fundamentals of statistical and thermal physics*. McGraw-Hill Book Company, New York, (1965).

[11] E. Altman and E. Demler. Condensed-matter physics: Relaxation after a tight squeeze. *Nature*, 449:296–297, (2007).

[12] R. Gati, B. Hemmerling, J. Fölling, M. Albiez, and M. Oberthaler. Noise thermometry with two weakly coupled Bose-Einstein condensates. *Phys. Rev. Lett.*, 96:130404, (2006).

[13] L. Boltzmann. *Vorlesungen über Gastheorie*. J. A. Barth, (1898).

[14] D. Ya. Ptrine. *Stochastic Dynamics and Boltzmann Hierarchy*. de Gruyter Expositions in Mathematics 48, (2009).

[15] C. J. Pethik and H. Smith. *Bose-Einstein condensation in dilute gases*. Cambridge University Press, (2001).

[16] S. Stringari and L. Pitaevskii. *Bose-Einstein condensation*. Oxford Science Publications, Great Clarendon Street, Oxford, (2003).

[17] C. W. Gardiner, P. Zoller, R. J. Ballagh, and M. J. Davis. Kinetics of Bose-Einstein condensation in a trap. *Phys. Rev. Lett.*, 79:1793, (1997).

[18] C. W. Gardiner and P. Zoller. Quantum kinetic theory. III. Quantum kinetic master equation for strongly condensed trapped systems. *Phys. Rev. A*, 58:536, (1998).

[19] C. W. Gardiner. *Handbook of stochastic methods*. Springer Verlag, (2004).

[20] C. Cohen-Tannoudji, J. Dupont-Roc, and G. Grynberg. *Processus d'interaction entre photons et atomes*. Savoirs Actuels, Editions du CNRS, (1996).

[21] H. P. Breuer. *The theory of open quantum systems*. Clarendon Press, (2006).

[22] B. N. Taylor P. J. Mohr and D. B. Newell. CODATA recommended values of the fundamental physical constants: 2006. *Rev. Mod. Phys.*, 80:633, (2008).

[23] M. H. Anderson, J. R. Ensher, M. R. Matthews, C. E. Wiemann, and E. A. Cornell. Observation of Bose-Einstein condensation in a dilute atomic vapor. *Science*, 269:198–201, (1995).

[24] K. B. Davis, M. O. Mewes, M. R. Andrews, M. O. Mewes, N. J. van Druten, D. S. Durfee, D. M. Kurn, and W. Ketterle. Bose-Einstein condensation in a gas of sodium atoms. *Phys. Rev. Lett.*, 75:3969, (1995).

[25] C. C. Bradley, C. A. Sackett, J. J. Tollett, and R. G. Hullet. Evidence of Bose-Einstein Condensation in an Atomic Gas with Attractive Interactions. *Phys. Rev. Lett.*, 75:1687, (1995).

[26] W. Ketterle. *Bose-Einstein condensation*. McGraw-Hill Encyclopedia of Science and Technology, 9th Edition, in print., (2009).

[27] V. V. Kocharovsky, V. V. Kocharovsky, M. Holthaus, C. H. R. Ooi, A. Svidzinsky, W. Ketterle, and M. O. Scully. Fluctuations in ideal and interacting Bose-Einstein Condensates. *Advances in Atomic, Molecular and Optical Physics*, 291:53, (2006).

[28] A. Einstein. Quantentheorie des einatomigen idealen Gases: Zweite Abhandlung. *Sitzungsber. Kgl. Akad. Wiss.*, 1925, (1925).

[29] A. Einstein. Quantentheorie des einatomigen idealen Gases. *Sitzungsber. Kgl. Akad. Wiss.*, 1924, (1924).

[30] S. N. Bose. Planck's law and the light quantum hypothesis. *Z. Phys.*, 26:178, (1924).

[31] O. Theimer and B. Ram. The beginning of quantum statistics: A translation of Planck's law and the light quantum hypothesis. *Am. J. Phys.*, 44:1056, (1976).

[32] M. R. Andrews, M. O. Mewes, N. J. van Druten, D. S. Durfee, D. M. Kurn, and W. Ketterle. Direct, non-destructive observation of a Bose-Einstein condensate. *Science*, 273:84–87, (1996).

[33] F. Sols and A. Leggett. On the concept of spontaneously broken gauge symmetry in condensed matter physics. *Found. of Phys.*, 21:353–364, (1991).

[34] F. Bardou, J. P. Bouchaud, A. Aspect, and C. Cohen-Tannoudji. *Lévy Statistics and Laser Cooling*. Cambridge University Press, (2000).

[35] K. B. Davis, M. O. Mewes, M. A. Joffe, M. R. Andrews, and W. Ketterle. Evaporative cooling of sodium atoms. *Phys. Rev. Lett.*, 74:5202, (1995).

[36] K. B. Davis, M. O. Mewes, M. A. Joffe, M. R. Andrews, and W. Ketterle. Evaporative cooling of sodium atoms. *Phys. Rev. Lett.*, 75:2909, (1995).

[37] C. C. Bradley, C. A. Sackett, J. J. Tollett, and R. G. Hulet. Evidence of Bose-Einstein condensation in an atomic gas with attractive interactions. *Phys. Rev. Lett.*, 75:1687, (1995).

[38] P. W. Anderson. Considerations on the flow of superfluid helium. *Rev. Mod. Phys.*, 38:298, (1966).

[39] A. Browaeys, A. Robert, O. Sirjean, J. Poupard, S. Nowak, D. Boiron, C. I. Westbrook, and A. Aspect. Thermalization of magnetically trapped metastable helium. *Phys. Rev. A*, 64:034703, (2001).

[40] M. Abramowitz and I. A. Stegun. *Pocketbook of mathematical functions*. Verlag Harri Deutsch, (1984).

[41] B. Kahn and G. E. Uhlenbeck. On the theory of condensation. *Physica (Amsterdam)*, 5:399, (1938).

[42] M. Holthaus, E. Kalinowksi, and K. Kirsten. Condensate fluctuations in trapped Bose gases: Canonical vs microcanonical ensemble. *Ann. Phys. (N. Y.)*, 270:198, (1998).

[43] T. P. Meyrath J. L. Hanssen G. N. Price C. S. Chuu, F. Schreck and M. G. Raizen. Direct observation of sub-poissonian number statistics in a degenerate Bose gas. *Phys. Rev. Lett.*, 95:260403, (2005).

[44] O. Penrose and L. Onsager. Bose-Einstein condensation and liquid helium. *Phys. Rev.*, 104:576–584, (1956).

[45] R. Kaiser, C. Westbrook, and F. David. *Lecture notes of Les Houches summer school, Session LXXII, Coherent atomic matter waves*. Springer, Paris, (1999).

[46] A. Griffin. Conserving and gapless approximations for an inhomogeneous Bose gas at finite temperatures. *Phys. Rev. B*, 53:9341, (1996).

[47] T. Nikuni and A. Griffin. Temperature-dependent relaxation times in a trapped Bose-condensed gas. *Phys. Rev. A*, 65:011601, (2002).

[48] Y. Castin and R. Dum. Low-temperature Bose-Einstein condensates in time-dependent traps: Beyond the U(1)-symmetry-breaking approach. *Phys. Rev. A*, 57:3008, (1998).

[49] C. W. Gardiner. Particle-number-conserving Bogoliubov method which demonstrates the validity of the time-dependent Gross-Pitaevskii equation for a highly condensed Bose gas. *Phys. Rev. A*, 56:1414, (1997).

[50] D. M. Stamper-Kurn, M. R. Andrews, A. P. Chikkatur, S. Inouye, H. J. Miesner, J. Stenger, and W. Ketterle. Optical confinement of a Bose-Einstein condensate. *Phys. Rev. Lett.*, 80:2027–2030, (1998).

[51] R. Walser, J. Williams, J. Cooper, and M. Holland. Quantum kinetic theory for a condensed bosonic gas. *Phys. Rev. A*, 59:3878, (1999).

[52] M. Reed and B. Simon. *Scattering theory*. Academic Press, inc., (1979).

[53] I. Bloch, J. Dalibard, and W. Zwerger. Many-body physics with ultracold gases. *Rev. Mod. Phys.*, 80:885, (2008).

[54] T. Mayer-Kuckuck. *Kernphysik.* Teubner, (2002).

[55] O. Hitmair. *Lehrbuch der Quantentheorie.* Karl Thiemig, (1972).

[56] H. Goldstein. *Classical mechanics.* Addision-Wesley Publishing Company, (1980).

[57] R. F. Streater. *Statistical Dynamics.* Imperial College Press, Singapore, (1995).

[58] C. Cohen-Tannoudji, J. Dupont-Roc, and G. Grynberg. *Photons and Atoms.* Wiley-VCH, (2004).

[59] M. Holland, J. Williams, and J. Cooper. Bose-Einstein condensation: Kinetic evolution obtained from simulated trajectories. *Phys. Rev. A*, 55:3670, (1997).

[60] E. Levich and V. Yakhot. Time evolution of a Bose system passing through the critical point. *Phys. Rev. B*, 15:243, (1976).

[61] B. V. Svistunov. . *J. Moscow. Phys. Soc.*, 1:373, (1991).

[62] Yu. M. Kagan, B. V. Svistunov, and G. V. Shlyapnikov. Bose-Einstein condensation in trapped atomic gases. *Phys. Rev. Lett.*, 76:2670, (1996).

[63] R. Walser, J. Williams, J. Cooper, and M. Holland. Quantum kinetic theory for condensed bosonic gases. *Phys. Rev. A*, 59:3878, (1999).

[64] H. J. Miesner, D. M. Stamper-Kurn, M. R. Andrews, D. S. Durfee, S. Inouye, and W. Ketterle. Bosonic stimulation in the formation of a Bose-Einstein condensate. *Science*, 273:1005–1007, (1998).

[65] R. Walser, J. Williams, J. Cooper, and M. Holland. Memory effects and conservation laws in the quantum kinetic evolution of a dilute Bose gas. *Phys. Rev. A*, 66:043618, (2002).

[66] H. T. Stoof. Nucleation of Bose-Einstein condensation. *Phys. Rev. A*, 45:8398, (1992).

[67] H. T. C. Stoof. Formation of the condensate in a dilute Bose gas. *Phys. Rev. Lett.*, 66:3148, (1991).

[68] H. T. C. Stoof. Initial stages of Bose-Einstein condensation. *Phys. Rev. Lett.*, 78:768, (1996).

[69] C. W. Gardiner and P. Zoller. Quantum kinetic theory: A quantum kinetic master equation for condensation of a weakly interacting Bose gas without a trapping potential. *Phys. Rev. A*, 55:2902, (1997).

[70] D. Jaksch, P. Zoller, and C. W. Gardiner. Quantum kinetic theory. II. Simulation of the quantum Boltzmann master equation. *Phys. Rev. A*, 56:575, (1997).

[71] D. Jaksch, C. W. Gardiner, K. M. Gheri, and P. Zoller. Quantum kinetic theory. IV. Intensity and amplitude fluctuations of a Bose-Einstein condensate at finite temperature including trap loss. *Phys. Rev. A*, 58:1450, (1998).

[72] D. Jaksch, C. W. Gardiner, K. M. Gheri, and P. Zoller. Quantum kinetic master equation for mutual interaction of condensate and noncondensate. *Phys. Rev. A*, 61:033601, (2000).

[73] C. W. Gardiner, M. D. Lee, R. J. Ballagh, M. J. Davis, and P. Zoller. Quantum kinetic theory of condensate growth: Comparison of experiment and theory. *Phys. Rev. Lett.*, 81:5266, (1998).

[74] M. D. Lee and C. W. Gardiner. Quantum kinetic theory VI: The growth of a Bose-Einstein condensate. *Phys. Rev. A*, 62:033606, (2000).

[75] M. Köhl, M. J. Davis, C. W. Gardiner, T. W. Hänsch, and T. Esslinger. Growth of Bose-Einstein condensates from the thermal vapor. *Phys. Rev. Lett.*, 88:080402, (1998).

[76] M. J. Davis, C. W. Gardiner, and R. J. Ballagh. Quantum kinetic theory VII: The influence of vapor dynamics on condensate growth. *Phys. Rev. A*, 62:063608, (2000).

[77] C. W. Gardiner and P. Zoller. *Quantum noise*. Springer Verlag, (2004).

[78] M. R. Andrews, M.-O. Mewes, N. J. van Druten, D. S. Durfee, D. M. Kurn, and W. Ketterle. Direct, nondestructive observation of a Bose Condensate. *Science*, 273:84, (1996).

[79] H. T. C. Stoof. Nucleation of Bose-Einstein condensation. *Phys. Rev. A*, 45:8398, (1992).

[80] W. Ketterle and N. J. van Druten. Bose-Einstein condensation of a finite number of particles trapped in one or three dimensions. *Phys. Rev. A*, 54:656, (1996).

[81] O. Penrose and L. Onsager. Bose-Einstein condensation and liquid helium. *Phys. Rev.*, 104:576–584, (1956).

[82] T. Nikuni, E. Zaremba, and A. Griffin. Two-fluid dynamics for a Bose-Einstein condensate out of local equilibrium with the non-condensate. *Phys. Rev. Lett.*, 83:10, (1999).

[83] R. Alicki. Field-theoretical methods. *Lecture notes at Institute of Theoretical Physics and Astrophysics, Poland*, March 20, (2006).

[84] N. N. Bogolyubov and N. N. Bogolyubov Jr. *Introduction to quantum statistical mechanics*. World Scientific Publishing, (1982).

[85] G. Lindblad. On the generators of quantum dynamical semigroups. *Commun. Math. Phys.*, 48:119, (1976).

[86] W. E. Lamb. Fine structure of the hydrogen atom by a microwave method. *Phys. Rev.*, 72:241, (1947).

[87] M. Sargent, M. Scully, and W. E. Lamb. *Laser physics*. Addison-Wesley Publishing Company (1974), Addison-Wesley Publishing Company, (1974).

[88] M. Giradeau and R. Arnowitt. Theory of many-boson systems: Pair theory. *Phys. Rev.*, 113:755, (1959).

[89] Yu. M. Kagan, B. V. Svistunov, and G. V. Shlyapnikov. . *Sov. Phys. JETP*, 75:387, (1992).

[90] S. Ritter, A. Öttl, T. Donner, T. Bourdel, M. Köhl, and T. Esslinger. Evaporative cooling of sodium atoms. *Phys. Rev. Lett.*, 98:090402, (2007).

[91] M. Köhl, M. J. Davis, C. W. Gardiner, T. W. Hänsch, and T. Esslinger. Growth of Bose-Einstein condensates from thermal vapor. *Phys. Rev. Lett.*, 88:080402, (2002).

[92] F. Gerbier, J. H. Thywissen, S. Richard, M. Hugbart, P. Bouyer, and A. Aspect. Experimental study of the thermodynamics of an interacting trapped Bose-Einstein condensed gas. *Phys. Rev. A*, 70:013607, (2004).

[93] Roy J. Glauber. *Quantum theory of optical coherence: Selected papers and lectures*. Wiley-VCH, Addison-Wesley Publishing Company, (2007).

[94] G. C. Wick. The Evaluation of the collision matrix. *Phys. Rev.*, 80:268, (1950).

[95] E. A. Burt, R. W. Ghrist, C. J. Myatt, M. J. Holland, E. A. Cornell, and C. E. Wiemann. Coherence, correlations, and collisions: What one learns about Bose-Einstein condensates from their decay. *Phys. Rev. Lett.*, 79:337, (1997).

[96] I. S. Gradsteyn and I. M. Ryzhik. *Table of integrals, series, and products*. Academic Press, (1995).

[97] M. R. Matthews, B. P. Anderson, P. C. Haljan, D. S. Hall, M. J. Holland, J. E. Williams, C. E. Wieman, and E. A. Cornell. Watching a superfluid untwist itself: Recurrence of Rabi oscillations in a Bose-Einstein condensate. *Phys. Rev. Lett.*, 83:3358, (1999).

[98] C. J. Myatt, E. A. Burt, R. W. Ghrist, E. A. Cornell, and C. E. Wieman. Production of two overlapping Bose-Einstein condensates by sympathetic cooling. *Phys. Rev. Lett.*, 78:586, (1999).

[99] M. Hugbart, J. A. Retter, A. F. Varón, P. Bouyer, A. Aspect, and M. J. Davis. Population and phase coherence during the growth of an elongated Bose-Einstein condensate. *Phys. Rev. A*, 75:011602, (2007).

[100] J. Esteve, J. B. Trebbia, T. Schumm, A. Aspect, C. I. Westbrook, and I. Bouchoule. Observations of density fluctuations in an elongated Bose gas: Ideal gas and quasi-condensate regimes. *Phys. Rev. Lett.*, 96:130402, (2006).

[101] I. Carusotto and Y. Castin. Condensate statistics in one-dimensional interacting Bose gases: Exact results. *Phys. Rev. Lett.*, 90:030401, (2003).

[102] E. Haller, M. Gustavsson, M. J. Mark, J. G. Danzl, R. Hart, G. Pupillo, and H. Ch. Nägerl. Realization of an Excited, Strongly Correlated Quantum Gas Phase. *Science*, 4:1224–1227, (2009).

Die VDM Verlagsservicegesellschaft sucht für wissenschaftliche Verlage abgeschlossene und herausragende

## Dissertationen, Habilitationen, Diplomarbeiten, Master Theses, Magisterarbeiten usw.

für die kostenlose Publikation als Fachbuch.

Sie verfügen über eine Arbeit, die hohen inhaltlichen und formalen Ansprüchen genügt, und haben Interesse an einer honorarvergüteten Publikation?

Dann senden Sie bitte erste Informationen über sich und Ihre Arbeit per Email an *info@vdm-vsg.de*.

**Sie erhalten kurzfristig unser Feedback!**

VDM Verlagsservicegesellschaft mbH
Dudweiler Landstr. 99         Telefon  +49 681 3720 174
D - 66123 Saarbrücken         Fax      +49 681 3720 1749
**www.vdm-vsg.de**

Die VDM Verlagsservicegesellschaft mbH vertritt

Printed by Books on Demand GmbH, Norderstedt / Germany